Moments in the life of a scientist

Bruno Rossi; a recent photo by Steve Borack.

MOMENTS IN THE LIFE
OF A SCIENTIST

Bruno Rossi
Massachusetts Institute of Technology

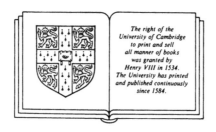

The right of the
University of Cambridge
to print and sell
all manner of books
was granted by
Henry VIII in 1534.
The University has printed
and published continuously
since 1584.

CAMBRIDGE UNIVERSITY PRESS
Cambridge
New York   Port Chester   Melbourne   Sydney

CAMBRIDGE UNIVERSITY PRESS
Cambridge, New York, Melbourne, Madrid, Cape Town, Singapore, São Paulo

Cambridge University Press
The Edinburgh Building, Cambridge CB2 8RU, UK

Published in the United States of America by Cambridge University Press, New York

www.cambridge.org
Information on this title: www.cambridge.org/9780521364393

© Cambridge University Press 1990

First published 1990
This digitally printed version 2008

A catalogue record for this publication is available from the British Library

Library of Congress Cataloguing in Publication data
Rossi, Bruno Benedetto, 1905–
[Momenti nella vita di uno scienziato. English]
Moments in the life of a scientist/Bruno Rossi.
p.   cm.
Translation of: Momenti nella vita di uno scienziato.
ISBN 0 521 36439 6
1. Rossi, Bruno Bendetto, 1905–   2. Physicists—Italy—Biography.
I. Title.
QC16.R495R6713   1990
530′.092–dc20   89–22283   CIP

ISBN 978-0-521-36439-3 hardback
ISBN 978-0-521-07015-7 paperback

To Florence, Frank, Linda –

Considerate la vostra semenza
fatti non foste a viver come bruti
ma per seguir virtute e conoscenza.

(Dante, *Inferno* XXVI, 118–20)

CONTENTS

PLATES

Between pp. 80 and 81

FOREWORD

Against the harmonious backgrounds assembled of green gardens and carved stone that the princes and the people of Italy have made in their cities, in Venice, Padua, Bologna, Florence, and Rome, these pages tell a vivid tale. Here is the rise of a new astronomy, the terror of tyrants and its harsh remedy in war, and the spread of the new ideas and tools of the devoted investigator to minds and lands far from the workshops of its origin.

You have read something like this somewhere before? A reader with any acquaintance with the history of science will catch the echo. But this is not another history of the seventeenth century. This is our time, our century, with its own ironies of wonder and of fear. Bruno Rossi is entirely a modern, as adept at electronics and quantum theory as he is in clear thought and sharp phrase, a man just as well known in New Mexico, Japan, Bolivia, or India as in the Tuscany of his first researches.

A young man eager to touch with newly-trained hand the mainspring of the world, he was among the few to found a new astronomy that displays an invisible part of the cosmos now as clear as the moons of Jupiter, and much more universal. It was in the years 1930 and 1931 that Rossi and his peers understood and first demonstrated that the mysterious radiation from on high, what we now call cosmic rays, was in fact particulate.

His telescopes were not fitted with glass lenses to augment the eye's grasp; they were metal tubes of gas whose signals were electrical pulses, invisible but rigorously countable if you had the right ingenious circuits. (The best-known of Rossi's circuit designs are still in use, though realized now in tiny silicon chips instead of with the glowing vacuum tubes of their beginnings.) Step by well-supported step, first one, two, three . . . then whole arrays of

Geiger counters and heavy shields artfully deployed, they marked
out the necessary path of the particles, not by images but merely by
recording simultaneous pulses – and their absence – to lead logically
to firm conclusions.

A new astronomy of charged particles and their secondaries was
born. The particles flew in, not on straight lines but on long tortuous
paths, from somewhere in distant space – where? – to the earth's
surface. It was soon proved that the particles were so penetrating, so
energetic, that their like had never been seen in the physics labs with
radioactive sources at hand.

During the rise of the first astronomy, the telescopic one, its home
Tuscany was at peace, though the stresses that brought the War of
Thirty Years across Germany were already distorting life on both
sides of the Alps. In the twentieth century the German wars were
more closely spaced. A lifetime could not escape one or the other.
Peaceable folk like the Rossis had perforce to leave behind the
brand-new lab at Padua that Bruno had designed, to seek refuge
overseas.

In a few years, the war had crossed to America. It is no surprise
that Bruno Rossi soon found himself at Los Alamos, his mastery of
sub-microsecond timing directed to development and test of the
atomic bomb. A faint trace of the record that his more than
lightning-fast detector took at the first desert test shows the signal
rising, rising as the chain reaction grows exponentially toward
fateful maturity.

The postwar years found the cosmic ray physicists for a while
happy monopolists of high-energy particle physics. It was then an
opportunistic, even a serendipitous science, based on the chance
infall of Nature's puzzling beam from afar. Cosmic rays were
sought worldwide, up on mountain tops or in stratosphere balloons
to look at the incoming beam, down in deep mines to filter out all
but the most penetrating rays, under the open sky to strew a score of
big counters over miles of prairie or forest land, the better to catch a
great disc of speeding particles, the billion-fold progeny of one rare
entrant particle of enormous energy. The hunt was open to players
in rich countries and in poor. By the early fifties, the romantic days
were mainly past. The time of the big particle accelerators had
come. As a rarity, some single cosmic particle still carries much
more energy than any you find in the latest magnetic tunnel, but for

most purposes particle research has left mountain and mine for the great national and international labs.

The strong, well-controlled beams within the big labs are fine for the study of particles, but helpless to plumb the sky. Bruno became more and more of an astronomer, albeit an active intervenor in space, and less of a laboratory physicist. He gives a neat chronicle of the second new astronomy that he and his colleagues (and some friendly competitors, too) found up there: an astronomy not of particles, nor of light nor of radio, but of X-rays (and gamma rays, too). Out where those rays come from there are strange stars and disturbed galaxies, spouting, spinning, and whirling, a turmoil little seen here in our quieter home, the solar system.

Rossi tells all this in his lucid, warm and personal way. Perhaps it is the growth of a worldwide community of Rossi students, most of them walking on the stage within these pages, that best stands for those times. Yet growth continues into these times when satellite probes are important tools of the physicist. Rossi led this quest too, out into interplanetary space, there to sound the plasmas that blow from the sun to modulate the cosmic rays, to tickle the comet's tails, and to disturb the upper atmosphere and magnetism of the earth.

It has been a glorious time for cosmic rays and for those who puzzled over them, on paper or with intrepid searches. After all these years, an elderly physicist takes pleasure in recalling the years of the Rossi curve, when a new kind of digital logic, based on banks of counters, not only built but compelled a student's belief in extraordinary new physics, demonstrated *alla Rossi.*

These pages delight us with an account of how it all felt, a portrait of the artist as a whole man. It is even more satisfying that Nora Rossi adds her keenly-observed and cosmopolitan chapter, based on a marriage that has celebrated its happy golden anniversary. She is able to explain to a clever stranger watching Bruno Rossi from the next table, once the man had conjectured that Bruno must either be poet or astronomer, that he was both!

The physicists knew it too, and the fascinated reader is likely soon to agree.

Philip Morrison
Cambridge, Massachusetts

PREFACE

At this time, when recent developments have brought science to a prominent position both in the cultural life and in the everyday life of our society, it is of interest to recall the activities of scientists in the years when the foundations of these truly revolutionary developments were laid.

I was one of these scientists.

When I look back to the pattern of my work, I recognize in it a restlessness which made me move from one problem to another, always hoping to detect some yet unknown aspect of Nature. This fascination for the secrets of Nature is the reason why I have been an experimental physicist. It is the reason why, for me, the most exhilarating moments have been those when an experiment gave a result contrary to all predictions, thus proving that the riches of Nature far exceed the imagination of Man.

Many persons lent their support in the preparation of these autobiographical notes. My daughter Linda and my wife Nora read the entire draft and were generous of constructive criticisms. My colleague in the Linguistics Department, Wayne O'Neil, polished the manuscript. Several of my present and past associates shared their recollections with me and offered comments and suggestions; among them Herbert Bridge, Benjamin Diven, John Linsley, George Clark, Martin Annis, Walter Lewin, Stanislaw Olbert and Claude Canizares. To all, my warmest thanks.

Prehistory

I do not remember when my interest in science began. Perhaps there was not a beginning. Perhaps it evolved gradually from seeds which already existed in my childhood, in the form of a curiosity for the everyday wonders of nature – a blossoming plant, the small treasures on the beach left behind by the receding tide. Clearer evidence of a scientific disposition would come when later I began looking for regularities, for relations of cause and effect and when I became conscious of the world which hides behind the world of our senses, and which, of this world, controls the behavior.

Even if what I say is not mere fantasy and even if there actually was in my nature an inborn tendency toward science, I know for certain that it was the influence of my father which turned this tendency into a lifelong commitment. My father was an electrical engineer. He had chosen this profitable profession because his family was in need of financial help following the unfortunate outcome of a business venture to which my grandfather had incautiously committed most of his personal fortune. If it had not been for this, he would certainly have become a scientist himself, for science was his great love. He often talked to me about one or another scientific news item that had aroused his interest. But it was not so much his words as my awareness of his regret for an unfulfilled dream that made me feel the fascination of science at a time when I did not yet appreciate its value.

My father was born in Bologna, my mother in Ferrara. But, soon after their marriage, my parents had moved to Venice, where the electrification of the city was under way. Participation in this very important project was, for my father, the beginning of a successful career.

And so it happened that I was born in Venice, where I spent most of the years of my youth. To Venice I still feel attached by the fond memories of those years. The most ancient memory, still strangely alive in my mind, goes back to my early childhood. I could not have been more than four years old. With my parents I was returning from Ferrara where we had gone to visit my mother's family. At the railway station, the train would stop, discharge the passengers and then move back. This meant that *the world ended in Venice.* Undoubtedly, other facts, other images had contributed in forming in my mind the idea that Venice was at the end of the world. Venice was surrounded by its *Laguna*; beyond the *Laguna* there was the long and narrow island of the Lido; beyond the Lido there was the sea; and beyond the sea nothing could be seen, nothing existed.

Also I clearly remember a childish illusion stemming from the peculiarities of my home town. Venice with its *calli*, with its canals, with its palaces, was something unique. Ferrara, Padua, other places I had seen were not cities; Venice was the only city in the world, and as such deserved a special position.

Of the memories of my adolescence, those that most often come back are of the long walks along the sea shore of the Lido deserted, in the early hours of the morning. I breathed the pungent air of the sea, I listened to the splashing of the waves, I let the wind blow in my face. The sun was still low on the horizon, and I was looking at the luminous trail produced by its reflection on the rippled surface of the water. (Was I perhaps wondering how to explain what I saw?)

Another recurrent memory is that of the rare, perfectly clear winter mornings when the air was so unusually transparent that the Alps surrounding Venice became clearly visible and appeared incredibly close (*Fata Morgana* if you are a child or a poet, anomalous atmospheric refraction if you are a scientist). On those mornings I would try to find a *sandalo* (a small gondola) and, accompanied by a friend, I would row standing on the stern, Venetian style, from the Lido, where I lived, to the school in Venice, where I studied. Venice, floating over the *Laguna* and the *Laguna* enclosed by the majestic circle of the Alps; the vision of a fairyland.

If Venice is my native town, Florence is the city where I was born as a physicist, where I became used to the hard work, and where I experienced the first disappointments and the first successes, which are the life of the scientist. I arrived in Florence in early 1928, with a position of assistant to the professor of experimental physics at the University. I had studied for two years at the University of Padua; then for another two years at the University of Bologna, where, in 1927, I had been awarded a doctoral degree in physics.

At that time, both the Universities of Padua and Bologna had in their faculties world-famous mathematicians, such as Gregorio Ricci-Curbastro in Padua, and Salvatore Pincherle in Bologna. I attended their lectures and was so impressed as to be almost tempted to become a mathematician myself.

Much less inspiring was my contact with physics during my university years. Perhaps my memory is failing me, but I remember having learnt little from the courses of general physics either in Padua or in Bologna. I do remember having followed with interest and profit a course in theoretical mechanics (*Meccanica Razionale*) by Professor Ernesto Laura in Padua and a course in theoretical physics by Professor Pietro Burgatti in Bologna. But the contents of these courses, again, were more mathematics than physics.

An important experience, however, was the long hours spent in the student laboratory of the physics institute of the University of Bologna. The laboratory was well equipped, and was used by only three students: two boys and one girl. The person in charge of the laboratory, Dr Orazio Specchia, had given us the key, and then had disappeared, leaving us free to do whatever we wanted – something for which I still am grateful. I took this opportunity to make a whole set of experiments on interference, diffraction and double refraction. I had thus learnt optics and had developed a special interest for this science, which years later, would induce me to write a textbook on optics.

But for an introduction to modern physics I had to wait for the arrival of Rita Brunetti. After spending several years as an assistant at the University of Florence, Ms Brunetti had joined the Bologna faculty as a professor. Despite the difference in age and in academic status, we had become friends. I am indebted to her for what she taught me, and for the introduction which brought me to Florence.

1

Arcetri

(1928–32)

The Physics Institute of the University of Florence rose, as it still does, among the olive trees of the hill of Arcetri, a short distance from the *Gioiello*, the villa where Galileo spent the last years of his life as a political exile, and a couple of miles from the seat of the University in downtown Florence. The director of the Institute, and professor of experimental physics, was Antonio Garbasso. As a young man, he had done some good work on topics of current interest, such as electromagnetic waves, X-rays and spectroscopy. But then World War I and the events of the post-war period had diverted his interest toward politics. He was now a senator of the Kingdom, and the mayor of Florence. However, he still found the time to go to Arcetri three times a week to deliver his lectures, and he still cherished the hope to see the laboratory, which he had built, become an important research center. Knowing that he could no longer actively conduct research himself, he had bestowed his still live interest in science on the young people working in his Institute. He took advantage of his influential position to support their work financially and took great pride in their achievements.

I remember my first meeting with Garbasso as if it had happened yesterday. I had just arrived in Florence and he received me in Palazzo Vecchio, in the handsome study assigned to the mayor of the city. He asked my age, 'almost 23 years' I answered, to which he commented: 'How can anyone be so young?'

In Arcetri I found a group of young people, all different in their personalities, in their attitudes, in their tastes, but united by bonds of friendship and by a similar commitment to science. There was Gilberto Bernardini, whose contagious enthusiasm brought color to our daily routine. He had arrived in Arcetri shortly before me with a doctoral degree from the University of Pisa. All the others were still students, but actually more collaborators and friends than pupils. Among the most faithful there was Giuseppe Occhialini, who already exhibited those qualities of imagination and originality that would characterize all of his future work. There was Daria Bocciarelli, whose presence greatly contributed to create that special atmosphere that was the 'spirit of Arcetri'. There were Guglielmo Righini, the future director of the Arcetri Observatory, and Beatrice Crinò, his future wife. There was Giulio Racah, a young scientist who was already internationally known as a promising theoretical physicist. And there was Lorenzo Emo Capodilista, whose courtesy and earnestness were, in our mind, the legacy of the old Venetian nobility to which he belonged.

We also had the pleasure and the benefit of the frequent visits of Enrico Persico, professor of theoretical physics at the University of Florence, who came up every week from his downtown office to disclose for us the mysteries of the new wave mechanics. (I still have the notes that Racah and I took of his lectures; masterpieces of clarity and precision.) We used to gather for lunch in a room of the *portincria* (the janitor's apartment) where, for a few *lire* the janitor's wife would satisfy our hunger with generous servings of spaghetti.

We were not the only inhabitants of the hill of Arcetri. Above our Institute rose the solar tower and the other structures of the Astronomical Observatory. We had close and friendly relations with the astronomers of the Observatory. The director, Professor Giorgio Abetti, was for us a friend and a protector, no less than he was for his own pupils. One of these, Attilio Colacevich, had become particularly attached to our group, of which he was a frequent welcome guest. Much appreciated was the initiative of Abetti to organize in a room of the Observatory a seminar which, once a week, brought together physicists and astronomers thus promoting exchange of ideas and personal contacts.

Life in Arcetri was rather austere. My monthly salary was 600 *lire* (about 30 dollars at the then-current rate of exchange). We never had money to buy heating fuel, and winters in Florence are often quite severe. As a protection against the cold, we would wear heavy woolen linings inside our smocks. The Institute was always late in paying its electric bills, and the only reason why electricity was not cut off was that the Institute's director was (or had been until recently) the mayor of the city.

I was eager to start working at some experimental project. But the ambition of the young scientist who does not yet see limits to his investigations did not allow me to settle for some run-of-the-mill program. My activity had to address itself to the fundamental problems of contemporary physics; it had to aim at the discovery of some secret of nature.

Month after month I pursued this dream. I remember that once, fascinated by the new horizons disclosed by Persico's lectures, Bernardini and I had decided to verify experimentally the predictions of wave mechanics by repeating, with slow electrons, one of the classical experiments of optical interference. On the face of it, this was a worthwhile experiment for, to our knowledge, no one had yet demonstrated directly the wave properties of matter predicted by wave mechanics. But we should have known that it was a project way above our possibilities (all we had to show for our efforts was a modest contribution to the study of the photographic effect of slow electrons). Not much more realistic, and equally doomed to failure, was an attempt to photograph the spectra of comet tails for the purpose of discovering their chemical composition. I was beginning to wonder whether my ambition had not led me into a blind alley and whether the time had not come to lower the aim. Fortunately I did not need to.

The initial period of my life as a would-be scientist, a period beset by uncertainties and saddened by disappointments, came to an end in the autumn of 1929, when I happened to read the historical article of the German physicists Walter Bothe and Walter Kohlhörster; *Das Wesen der Höhenstrahlung (The Nature of the Radiation from Above).*

Until then I had not been much interested in the *Höhenstrahlung*, or *cosmic radiation*, to use the suggestive expression introduced by Robert Millikan. I did not see how I could find material for research in any aspect of this phenomenon. By then, the celestial origin of cosmic radiation had been established beyond any reasonable doubt. The nature of the source was still a matter of speculation. Most widely advertised was a theory advanced by Millikan, according to which cosmic rays were the 'birth cry of atoms', being born in the form of gamma rays, from the energy set free in the synthesis of heavier atoms through fusion of primeval hydrogen atoms (in essence the same process which is exploited in the hydrogen bomb).

This was certainly a fascinating hypothesis, for it suggested an answer to two of the most fundamental problems of contemporary science: i.e., the origin of the atoms of the different elements and the origin of cosmic rays. Occhialini was fascinated by it and I remember that it was he who called my attention to the paper where Millikan had presented his ideas. But I was skeptical because, in my opinion, the interpretation of the experimental observations on which Millikan had built his theory was not convincing. On the other hand, I did not see how, being an experimentalist, I could contribute significantly to the current speculations about the origin of cosmic rays.

Lastly, there was the problem of the nature of cosmic rays. But this problem was not of much concern because, according to the general view, the answer was already known. The argument ran as follows: cosmic rays were known to have a penetrating power far exceeding that of any other known rays. Of these, the most penetrating were the gamma rays of some radioactive sources. Moreover, according to current theories, the penetrating power of gamma rays was expected to increase steadily with increasing energy. Therefore, the astonishingly penetrating cosmic rays could not be anything else but gamma rays of very great energy.

For the first time, the gamma-ray hypothesis was submitted to an experimental test of Bothe and Kohlhörster. In their experiment they used an instrument that had just been developed by the well known German physicist Hans Geiger and his pupil William Müller. I am referring to the *tube counter* later to become known as the *Geiger-Müller counter*, or *G.M. counter* for short. While the ionization chamber, used previously in cosmic ray studies, measures the total ionization of all particles which traverse its sensitive volume in a given time interval, the counter signals, with an electric pulse, the arrival of every single ionizing particle.

Sometime later Bothe was to tell me the story of the origin of the

Bothe and Kohlhörster experiment. It seems that Kohlhörster had been experimenting with tube counters in order to test their performance as cosmic-ray detectors. During his work, he noticed that two counters, placed one above the other, often produced simultaneous pulses. Puzzled, he told Bothe about this observation, and Bothe immediately suggested that the simultaneous pulses, the *coincidences*, as they become known, must be due to the passage of individual charged particles through both counters. (The operation of G.M. counters relies on the ionization produced in a gas by the incident particles; thus only charged particles which are capable of ionizing a gas, can be detected by counters.)

No charged particles capable of traversing two or three counter walls were known at that time. So Bothe concluded that the particles responsible for the coincidences must belong to cosmic radiation. By itself, the existence of ionizing particles in cosmic radiation did not contradict the gamma-ray hypothesis. It was known, in fact, that gamma rays do not ionize directly but do so through the intermediary of secondary ionizing particles which they produce in matter. At that time, the only known interaction of high-energy gamma rays with atoms was the so-called *Compton effect*, whereby photons knock electrons out of atoms and transfer part of their energy to them, thus giving rise to a secondary radiation consisting of fast-moving electrons. It was also known that these electrons, just because they lose energy at a fast rate by ionizing atoms, have a very small range in matter.

Bothe and Kohlhörster saw that the most direct line of attack to the problem of the nature of cosmic rays was to examine experimentally the properties of the particles detected by the counters, so as to decide whether or not these particles might be Compton electrons. The experimental arrangement used for this study (see Fig. 1.1) consisted essentially of two tube counters, placed horizontally one above the other, a few centimeters apart. Coincidences were recorded with and without a 4.1 cm-thick gold block between the counters. (Gold was chosen as an absorber because of its high density.) If the charged particles recorded without the gold block were Compton electrons, then even a thin absorber between the counters would have stopped all of them and suppressed the coincidences. Instead, it turned out that the insertion of the gold block produced only a small decrease in the coincidence

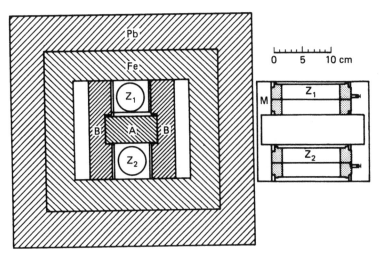

Fig. 1.1. The experiment of Bothe and Kolhörster.
Coincidences between counters Z_1 and Z_2 are produced by
cosmic-ray particles traversing both counters. Observations
were made both with and without a 4.1 cm thick gold absorber
between the counters. The coincidences observed with this
absorber demonstrate the existence of cosmic-ray particles
capable of traversing 4.1 cm of gold. (Gold was chosen as an
absorber because of its high density.) From an article by W.
Bothe and W. Kohlhörster in *Zeitschrift für Physik*, **36**, 751
(1929).

rate. This meant that most of the charged particles were capable of
traversing this block. It was hardly possible to identify these
particles with Compton electrons, and it was concluded that they
were not the secondary particles of a gamma radiation. Therefore
the cosmic radiation observed near sea level did not consist of
gamma rays, but rather of ionizing particles. Going further, Bothe
and Kohlhörster hypothesized that the primary cosmic radiation
(the radiation striking the Earth from outer space) was a stream of
charged particles and that the local cosmic radiation (the radiation
found at the place where the observations are made) consisted of the
particles of the primary radiation capable of traversing the over-
lying atmosphere.

We know today that this tentative assumption about the
nature of primary cosmic rays was essentially correct. We also
know, however, that the propagation of cosmic rays through the

atmosphere is a much more complex phenomenon than originally conceived.

The announcement of the discovery that the cosmic radiation is made of particles was received by the scientific community with great surprise and great interest, not free of doubts. I, for one, focussed my attention on the new problems created by this discovery. What, actually, were the particles found in cosmic rays? Were they particles of the same kind as those already known, only endowed with much greater energies? Or were they particles of a different, new kind? And, in either case, what were their physical properties? Here lay before me a field of inquiry rich in mystery and promises. Working in a field of this kind had been my dream. Now it seemed that this dream was coming true.

I started working immediately with the generous support of my friends in the group. And so began one of the most exhilarating periods of my life. Was it the excitement of the explorer who first ventures into an unknown land? Was it the companionship of people with whom I could share my enthusiasm, my hopes, my anxieties? Was it the subtle charm of the Tuscan countryside?

Others joined, and the work proceeded rapidly. In a few weeks our first counters were ready. At that time, the construction of a counter was nothing short of witchcraft. According to the German prescription, the cathode was to be a zinc tube. Zinc tubes were unavailable in Italy, so we had to build our own tubes by wrapping a zinc sheet around a metal cylinder and soldering the edges together. The anode was to be a thin steel wire, slightly oxidized by immersion in a solution of nitric acid. The wire was to be stretched along the axis of the tube and supported by two hard-rubber stoppers, which closed the ends of the tube and were made air-tight with some sort of wax. The tube was then evacuated through a thin glass pipe inserted in one of the stoppers, and then filled at $\frac{1}{10}$ of an atmosphere (atm) of pressure with dry, dust free air. The glass pipe was sealed by melting its walls on a flame, and the counter was taken off the filler, ready to be used.

When in use, the tube was brought to a negative electric potential on the order of 1000–1500 volts (V), and the wire was taken to ground through a resistor. The arrival of an ionizing particle would start a discharge, which would induce an electric pulse on the wire.

The ground resistor would promptly extinguish the discharge, making the counter ready to record the next ionizing particle. For this purpose a very high resistor was needed, amounting to more than a billion (10^9) ohms. No such resistors were commercially available, so that we had to prepare them ourselves by filling small glass tubes with some appropriate mixture of organic fluids (self-quenching counters were still in the future).

Bothe and Kohlhörster had recorded their coincidences by connecting the wires of the counters to two separate fiber electrometers, which were imaged on a moving photographic film. By some clever device, involving the use of a fast-oscillating screen, they had achieved a time resolution of $\frac{1}{100}$ of a second (which means that two unrelated pulses of the two counters would have been recorded as coincident if then they occurred within a time interval of less than $\frac{1}{100}$ of a second). I felt that the power of the coincidence method would be greatly enhanced if one could devise a method of recording coincidences that would be less cumbersome than that used by Bothe and Kohlhöster, that would achieve a better time resolution and, most importantly, that could be used to record coincidences between the pulses of more than two counters. And so the classical coincidence circuit was born, which was my first contribution to cosmic-ray research. For years to come, it was my main research tool, as well as that of many other experimenters. It is today the basic circuit of modern computers.

The circuit consisted of several triodes* with the grids coupled electrostatically to the wires of the counters and with the plates connected together by a wire which was brought to the positive terminal of an electric battery through a resistor. Simultaneous pulses of all counters would produce a signal, in the form of a sudden change in the electric potential of the plates. Fig. 1.2, which appeared in 1930 in a letter to *Physics Review*, shows the construction of the first coincidence circuit and explains in some detail its operation. With this circuit, the time resolution was approximately $\frac{1}{1000}$ of a second. It became better in later models. At first,

---

* Triode: a simple vacuum tube used in the early radio devices. It contains three electrodes: a filament heated by an electric current which emits electrons; a plate kept at a positive electric potential which collects these electrons; and a grid at a variable electric potential, located between the filament and the plate, which controls the flow of electrons.

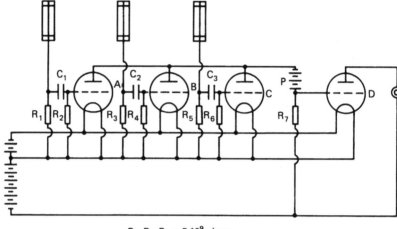

$R_1, R_3, R_5 = 5 \cdot 10^9$ ohms.

$R_2, R_4, R_6, R_7 = 8 \cdot 10^6$ ohms.

$C_1, C_2, C_3 = 10^{-4} \ \mu F.$

Fig. 1.2. The diagram of the original coincidence circuit published in 1930. The plates of the triodes (three in the figure) are connected together and brought to the positive terminal of a dry-cell battery (P), whose negative terminal is grounded through a resistor (R_7). The grids for the three triodes are coupled electrostatically to the wires of the G.M. counters.

In a quiescent state, the grids of the three triodes are at ground potential, causing a current to flow in the triodes and in the common plate resistor (R_7). The potential drop through this resistor holds the grid of the triode (D) at a negative potential, thus inhibiting the flow of plate current. The discharge of a counter will apply a negative potential to the grid of the corresponding triode, thus stopping its plate current. When this happens in one or two triodes the current flowing in the unaffected triode will still create a potential drop in the ground resistor (R_7) sufficient to inhibit the current in the triode (D). Only when a threefold coincidence occurs, i.e., when all three counters are discharged simultaneously will the current in the resistor (R_7) stop, bringing the grid of the triode (D) to ground potential and starting a current in this triode. From an article by B. Rossi in *Nature*, **125**, 636 (1930).

coincidences were recorded acoustically as clicks in a head phone; this meant that I or one of my collaborators had to permanently sit by the instrument. For all our dedication, we soon grew bored with this requirement. So the headphone was discarded and replaced by an electrometer whose deflections were recorded photographically. Eventually the electrometer was also discarded in favor of a mechanical counter.

For the record, I wish to add that, in the meantime, Bothe had also developed an electronic coincidence circuit. This circuit (which was based on the use of a different kind of vacuum tube, a penthode) was much less practical than mine. Moreover, it could only record twofold coincidences. As far as I know, I was the only one to use this circuit at the time.

Having in hand, with the coincidence circuit, a new technique which greatly expanded the possibilities of counters as tools for cosmic-ray research, I started immediately to venture into several preliminary experiments. Within a few months, I had measured the efficiency of the G.M. counters and its dependence on the negative voltage applied to the tubes. For this purpose I arranged three counters vertically one above the other and compared the frequency of threefold coincidences with the frequency of twofold coincidences between the uppermost and lowest counter. I found that with a sufficiently high voltage applied to the tubes, the efficiency was nearly one hundred percent, a result which placed a lower limit on the ionizing power of the cosmic-ray particles. This lower limit was of some interest at the time when nothing was known about the properties of cosmic-ray particles. I also verified that cosmic rays come preferentially from the vertical direction by comparing the frequency of the coincidences between two counters placed vertically one above the other with the frequency of the coincidences between the same counters placed one next to the other.

Finally I made an attempt to verify that cosmic-ray particles carried an electric charge, and to find the sign of this charge. For this purpose, I tried to detect the deflection that charged particles were supposed to undergo on traversing a bar of magnetized iron. Having failed to observe the effect I had anticipated, I thought that perhaps my arrangement was not sufficiently sensitive. I remember

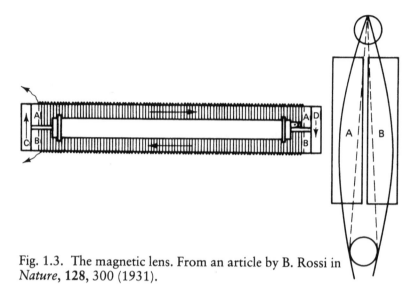

Fig. 1.3. The magnetic lens. From an article by B. Rossi in *Nature*, **128**, 300 (1931).

that during one of my frequent visits to the nearby University of Pisa, I went to see Antonio Puccianti, professor of physics at that University, and told him about my unsuccessful experiment. Puccianti listened, then came up with an interesting suggestion, which was to use a *magnetic lens*, i.e., a device that would concentrate particles of a given sign and defocus particles of the opposite sign. One could produce a magnetic lens by means of two iron bars, magnetized in opposite directions and placed one next to the other, with a counter above and one below them as shown in Fig. 1.3. Depending on the sign of the charge and on the direction of the magnetization, the particles traversing the upper counter would be focussed upon the lower counter or deflected away from it. An important feature of the suggested device was that the magnetic field lines close upon themselves within the iron, forming what is known as a *closed magnetic circuit*; with this particular configuration, it was easy to create a strong magnetization of the iron bars.

Back in Arcetri I constructed a magnetic lens and, using it, I attempted again to detect a magnetic deflection of cosmic-ray particles. I did observe an effect corresponding to positive particles, but the effect was not outside the experimental errors and at any rate, it was much smaller than I had anticipated. I was puzzled by this negative result, for it was not thinkable that cosmic-ray particles did not carry any electric charge. Of course, I did not know at that time that the cosmic radiation contains both positive and negative particles. Incidentally, several years later the magnetic lens found a useful application in some experiments by other scientists where it was used to study the behavior of positive and negative mesons.

In the meantime I had written to Bothe describing what I had been doing and telling him that I would like to spend a few months in his laboratory, at the Physikalische–Technische Reichsanstalt in Berlin–Charlottenburg. Bothe answered that he would be happy to receive me. Garbasso succeeded in obtaining a travel grant and so, in the early summer of 1930, I set out for Germany.

This was my first trip abroad. It was heart-warming to find Bothe waiting for me at the Berlin railway station. He accompanied me to a nice room which he had rented at a modest price in a working-class section of town not far from the Reichsanstalt. When I was rested, he invited me to his home and introduced me to his family in an effort to make me feel at home in his city.

The memory of that summer is still vivid in my mind. Berlin was then the heart of modern physics. For the weekly seminars the lecture room was crowded with physicists of all ages. The first row of benches looked like a hall of fame. Sitting there were scientists whose names were known to me as the creators of the new science. Unconsciously, I had felt that they hardly would look like ordinary human beings. And yet there they were, attentive and unassuming, Albert Einstein, Max Planck, Otto Hahn, Lise Meitner, Max Von Laue, Walter Nernst and Werner Heisenberg. Most of them worked at one or another of the several institutions in and around Berlin; others came from neighboring cities. Patrick Blackett was also there, visiting from England, and my friendship with him began on that occasion.

I soon developed a feeling of gratitude and warm friendship for Bothe. He had opened the doors of his laboratory for me and was always ready to support my work with his advice. I learned to fully appreciate his behavior toward me when I gradually realized that by nature he was not overly outgoing or trustful. In this connection I still remember a curious episode. I had just begun to work when I noticed that his G.M. counters were better than those I had built in Arcetri: they were more stable and the value of the potential applied to the tube was less critical. I was puzzled about this matter until one day Bothe took me aside with a mysterious air and began: 'I will tell you a secret, but you must promise not to give it away to anyone'. After I had promised, he continued: 'my counters do not have a steel wire, as advertised; they have an aluminum wire'. To my shame, I must confess that, upon my return to Italy, I was not able to keep the secret of the aluminum wire from my friends in Florence and Rome. But I relieved my conscience by requiring of them the same oath of secrecy that Bothe had required of me.

During my short visit to the Reichsanstalt I repeated, with some improvements, the experiment of Bothe and Kohlhöster. In my experiment I compared the rate of coincidences between two counters, placed one above the other, with a 9.7 cm-thick lead shield placed, alternately, above and between the counters. I argued that if the observed particles were actually primary particles that had traversed the atmosphere the coincidence rate should have been the same in both cases. Instead, I found that with the lead above the counters there was an excess of about four percent in the coincidence rate over that observed with the lead between the counters. I did not think that this result contradicted the interpretation of cosmic-ray phenomena proposed by Bothe and Kohlhöster, but was rather evidence for the production of secondary particles by cosmic rays in the lead shield.

Another result of my visit in Berlin was the beginning of my interest in the geomagnetic effects on cosmic rays, an interest from which was later born a major research program. But more about this later.

In the autumn I was back in Italy. The work in Arcetri was proceeding well, and receiving increased attention, so that by 1931, Italy could boast of two research centers active in two of the most important fields of contemporary physics. The first was the well-established center of Via Panisperna in Rome where, under the guidance of Enrico Fermi, the complex problems of the nuclear structure were being investigated. The second was the younger center of Arcetri in Florence, where work was centered on the problems of cosmic rays.

I can testify that there was never any rivalry between the scientists working in Rome and Florence, but rather there was a keen mutual interest in the achievements of the two groups, and a sense of comradery born of the feeling that they were both aiming, through different roads, for the same goal: which was to achieve an understanding of the still hidden aspects of nature. From this common purpose there naturally arose a bond of friendship, which lasts to this day. I grew particularly close to Fermi, whom I visited frequently in Rome. (We even wrote a joint paper dealing with the geomagnetic effects on cosmic rays.)

In the autumn of 1931 the Royal Italian Academy, with Guglielmo Marconi as president, sponsored a conference on nuclear physics which brought to Rome, from all over the world, the most illustrious scientists interested in this and related fields of physics. At the invitaiton of Fermi, I gave an introductory speech on the problems of cosmic rays. The opening statement clearly reflects the mystery still surrounding this phenomenon. I said:

> The most recent experiments have produced evidence of such strange events that we are led to ask ourselves whether the cosmic radiation is not something fundamentally different from all other known radiations; or, at least, whether in the transition from the energies which come into play in radioactive phenomena to the energies which come into play in cosmic-ray phenomena the behavior of particles and photons does not change much more drastically than until now it was possible to believe.

After presenting the experimental results which proved conclusively that the cosmic radiation reaches the Earth from outer space, I went on to examine the problem of the origin of this radiation. I discussed in some detail Millikan's assumption that cosmic rays are born from the synthesis of atoms, and explained the reasons why I thought that this assumption could not be correct.

My strongest argument was based on energy considerations. As stated by Millikan, the maximum energy released as gamma rays in the synthesis of atoms amounts to 447 electron volts (eV). This is clearly also the maximum energy of the charged particles which are produced in the interactions of primary cosmic rays with matter. On the other hand, my own experiments had shown that many of these particles have a range greater than 10 centimetres (cm) of lead; and the well-established theory of energy losses of charged particles in matter told us that particles with such a large range must have an energy much greater than the limit of 447 eV set by Millikan's theory.

Having disposed of Millikan's theory, I turned to the problem of the nature of the cosmic radiation, making a clear distinction between the primary radiation and the local radiation. On the basis of the experiment of Bothe and Kohlhörster and of my own subsequent work, I felt entitled to assert that the local cosmic radiation consisted of charged particles and that these particles were not produced locally by gamma rays.

I did not think that I could make an equally definite statement about the nature of the primary radiation. Two different theories were still in competition. According to the first theory, primary cosmic rays were high-energy charged particles. This theory was known as the *corpuscular theory*. According to the second theory, primary cosmic rays were high-energy gamma rays (although not necessarily the *birth-cry* of atoms). This theory was known as the *wave theory* (because gamma rays are very short electromagnetic waves). Both theories had some difficulties, which, however, did not appear insurmountable. So, while personally I favored the corpuscular theory, I felt in all fairness that I could not yet reject the wave theory outright.

My talk had a mixed reception. Millikan clearly resented having his beloved theory torn to pieces by a mere youth, so much so that from that moment on he refused to recognize my existence. (In retrospect I must admit that I might have been more tactful in my presentation.) On the other hand my talk roused the interest of Arthur Compton who previously had not worked on cosmic rays. Years later he was kind enough to tell me that his interest in cosmic rays was born from my presentation.

The Rome conference provided the first occasion for the proponents of the new corpuscular theory to present this theory to the scientific community, still strongly attached to the old wave theory. So this conference marked the beginning of the historical debate about the nature of cosmic rays, which was to continue for several years. In the United States, this debate was often bitter, involving the personal prestige of the scientists committed to opposite views. For some reason, this did not happen on the other side of the Atlantic. In fact, two of the European scientists who were among the most staunch supporters of the wave theory – Lise Meitner and Erik Regener – were also among my dearest friends and among the scientists for whom I had the greatest esteem.

The discussions at the Rome conference reinforced my opinion that further measurements of the penetrating power of cosmic-ray particles would produce important information about the nature, properties, and origin of these particles. As I already mentioned, from direct experiments I knew that most cosmic-ray particles had a range greater than 10 centimetres of lead. Moreover, by indirect arguments, I had become convinced that many particles must have much greater ranges. I thought that it was important to verify experimentally this conjecture, which, if correct, would have put to rest once and for all the gamma-ray theory.

I planned to do this experiment by the usual method of counting coincidences between counters arranged vertically one above the other and separated by absorbers of suitable thickness. In a first experiment, using just two counters in coincidence, I measured the absorption of cosmic-ray particles up to a thickness of 25 centimeters of lead. When planning to extend the measurements to absorbers of greater thickness, I realized that with two counters placed sufficiently far apart to allow these absorbers to be placed between them, the number of 'true' coincidences would have been smaller than the number of 'chance' concidences, i.e., of coincidences due to the almost simultaneous passage through the counters of two unrelated particles. I overcame this difficulty by inserting a third counter between the other two and recording threefold instead of twofold coincidences, thereby cutting down the frequency of chance coincidences to a negligible value (see Fig. 1.4).

The results of my measurements may be summarized as follows. The number of particles capable of traversing a given thickness of lead decreases fairly rapidly in the first 10 centimeters. Then, however, the rate of decrease with increasing absorber thickness becomes very slow, so that as much as one-half of the particles emerging for 10 centimeters of lead are able to penetrate 1 meter of this metal.

It is difficult today to appreciate how hard it was for the majority

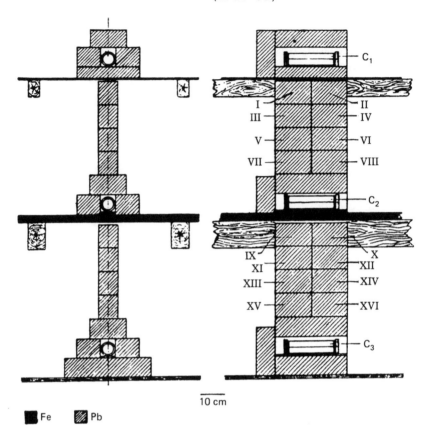

10 cm

■ Fe   ▨ Pb

Fig. 1.4. Experiment which proved that the cosmic radiation contains particles capable of traversing one meter of lead. These particles are recorded by the threefold coincidences between counters C_1, C_2 and C_3. From an article by B. Rossi in *Nature*, **20**, 65 (1932).

of the scientific community to accept this result. After all, the most penetrating ionizing particles known at that time (β-rays from radioactive sources) had ranges in lead of only a fraction of a millimeter. Doubts were raised about the validity of the coincidence method, and I had to perform a number of test experiments to dispel these doubts.

Striking as the discovery of particles endowed with such incredibly large penetrating power was, even more important, perhaps, was the implication of this discovery for the debate on the nature of the primary cosmic radiation. For according to the wave theory, cosmic-ray particles in the local radiation should have been Compton electrons produced by gamma rays somewhere in the atmosphere. By no stretch of the imagination could one suppose that such secondary electrons might have a range greater than 1 meter of lead.

In the discussions at the Rome conference, Bothe and I had reported the results of some experiments by several scientists (including my experiment in Berlin) which suggested the production of secondary particles by the interaction of cosmic-ray particles with matter. But the evidence for this effect was not very definite and, for the most part, it was of a rather indirect nature. Therefore, upon my return to Arcetri, I decided to attempt observing this phenomenon directly. For this purpose, I placed three counters in a triangular configuration so that a single particle travelling along a straight line could not traverse them all (see Fig. 1.5). With the counters enclosed in a lead box with walls a few centimeters thick I observed a large number of threefold coincidences. Removing the top part of the box caused the number of coincidences to decrease considerably. It was thus clear that most of the coincidences observed with the box closed were due to groups of associated particles (at least two) arising from interactions of cosmic rays in the box top. Qualitatively, this was the effect I had been looking for. But the size of the effect, as evidenced by the large number of coincidences, was astounding. It showed that cosmic rays were capable of producing an enormously more abundant secondary radiation than any other known rays. So incredible were my results that a German magazine (*Naturwissenschaften*, if I remember correctly) refused to publish my paper. This paper was later accepted by *Physikalische Zeitschrift*, after Heisenberg had vouched for my reliability.

I was greatly excited. Here was a new unexpected phenomenon, a surprising new property of the still mysterious cosmic rays. I continued to experiment using a variety of configurations for the counters, placing upon them, at different distances, layers of lead and of other materials, inserting absorbing shields in different positions. The most significant results obtained in this study are summarized in Fig. 1.6 which shows the dependence of the

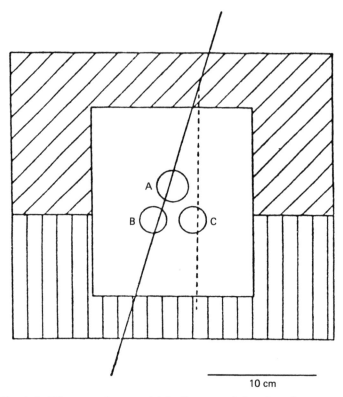

10 cm

Fig. 1.5. The experiment which discovered the abundant production of a secondary radiation by cosmic rays. Coincidences between G.M. counters A, B, and C are produced by groups of secondary particles generated by cosmic rays in the lead shield above the counters. From an article by B. Rossi in *Physikalische Zeitsch.*, **33**, 304 (1932).

coincidence rate between three counters in a triangular array on the thickness (in mass per unit area) of layers of lead and iron placed above them.

If we focus our attention upon the lead curves (I and II) we see that the rate of coincidences reaches a maximum between 10 and 20 g/cm^2, which shows that the secondary particles produced in lead have a range of this magnitude. Surprising was the observation that, beyond the maximum, the curves dropped much more rapidly than the known absorption curve of penetrating particles. This meant that the secondary interactions were not produced by the penetrat-

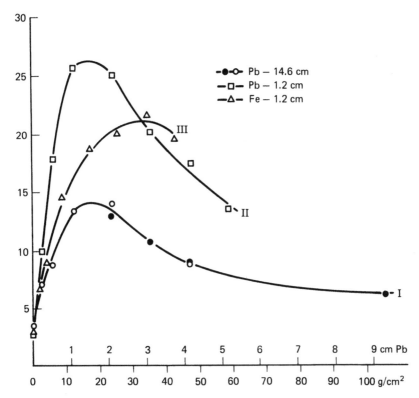

Fig. 1.6. The rate of coincidences between three counters in a triangular array (see Fig. 1.5) as a function of the thickness of a screen of lead or iron placed above the counters. Curves I and II refer to measurements taken with lead screens at distances of 14.6 and 1.2 cm, respectively, above the counters. Curve III refers to measurements taken with an iron screen at a distance of 1.2 cm above the counters. From an article by B. Rossi in *Zeitschrift für Physik*, **82**, 151 (1933).

ing particles (as I had every reason to expect), but by a different, much softer component of the cosmic radiation, whose existence was previously unknown.

I published my observations in 1933. The next important step toward an understanding of the phenomenon revealed by these observations was taken the following year by Blackett and Occhialini, using a newly developed technology: the *counter-controlled cloud chamber*.

Before I come to this, I must go back and recall that in the early thirties, to my knowledge, no one in Italy knew how to build or operate a cloud chamber. The best way to remedy this serious deficiency in the technological know-how of our scientific community was to arrange for some Italian physicist to learn the trade in a foreign laboratory. I thought that Occhialini would be the person best qualified for this assignment. Moreover, I knew that Occhialini himself was anxious to work abroad for a while. At that time, among European physicists, Blackett was regarded as the scientist most experienced in the technology and use of cloud chambers. On the other hand, Occhialini, working in Arcetri, had become familiar with the new coincidence technique, a technique virtually unknown in England. Thinking that Occhialini, in England, could provide as well as receive valuable technological information, I wrote to Blackett and with him, planned for Occhialini to spend some time working in his laboratory. And so, in 1931, Occhialini departed for Cambridge.

To describe precisely how a new scientific concept was born is always a difficult task; more so when there is reason to suspect an excessive degree of modesty on the part of the persons involved. My own recollection, for what it may be worth, is that, on leaving Arcetri, Occhialini already entertained the idea of pooling his own with Blackett's expertise for the purpose of building a counter-controlled cloud chamber, i.e., a cloud chamber whose expansions would be triggered by a coincidence arrangement signalling the arrival of cosmic-ray particles upon the chamber. The efficiency of such a device would have been much higher than that of a cloud chamber operated at random because the sensitive time, after each expansion, is only on the order of $\frac{1}{100}$ of a second, and the probability that, by chance, a cosmic-ray particle should traverse the chamber in this very short time interval is very small.

The realization of a counter-controlled cloud chamber was a difficult task, primarily because of the requirement that the chamber be expanded with an exceedingly short delay after a signal of the coincidence circuit. Eventually the joint efforts of Blackett and Occhialini overcame these difficulties, and in 1933 the counter-controlled cloud chamber was ready to operate. Among the first pictures obtained with this device many showed groups of associated particles which, as the authors pointed out, were undoubtedly

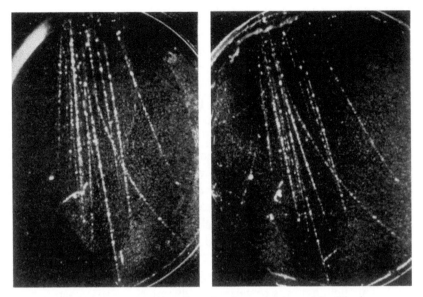

Fig. 1.7. Stereoscopic picture of a shower detected by Blackett and Occhialini with a counter-controlled cloud chamber, published in *Proceedings of the Royal Society*, **A149**, 699 (1933).

a manifestation of the same interactions which I had detected the year before with my counter experiments. Now, the observations with a cloud chamber made it possible to examine in detail the character of their interactions.

The most striking result was the great variety and the great complexity of the phenomenon. Some of the groups that were observed contained more than twenty particles. Some of the particles appeared to diverge from a single point, others from two or more points (Fig. 1.7). The groups of particles appearing in the cloud chamber pictures became known as *showers*. The chamber was operated in a magnetic field which often made it possible to identify the shower particles. It was thus found that at least most of them were *positive* or negative electrons.

The counter-controlled cloud chamber, while most valuable, did not detract from the usefulness of the counter technique because it was the only one that could be used in experiments requiring the observation of large numbers of interactions. Thus when the two physicists Homi Bhabha from India and Walter Heitler from Germany developed the theory of shower production they used the curve of the coincidence frequency versus lead thickness (shown in Fig. 1.6), a curve then known as the *Rossi Curve*, as an experimental test of their results.

Turning to a different subject, I recall that in the early 1930s, while the indirect evidence against the wave theory of cosmic rays was becoming increasingly compelling, the problem of the nature of the primary radiation – a problem of fundamental cosmological importance – had not yet been attacked by direct observations outside the atmosphere. Space vehicles capable of experimenting on the cosmic radiation before it enters the atmosphere were several decades in the future, and the technology of high-flying balloons was still not sufficiently advanced.

In principle, however, there was a possibility of obtaining information on primary cosmic rays by means of the Earth's magnetic field. This field extends to very large distances, so that cosmic rays will encounter it before collisions with atmospheric gasses may change their identity. If primary cosmic rays were gamma rays or other neutral particles, they would cross the Earth's magnetic field undisturbed. If, instead, they were electrically-charged particles, their trajectories would be deflected by the magnetic field and, consequently, their intensity distribution over the Earth's surface would be modified. The difficulty that scientists encountered in using the geomagnetic field in cosmic-ray studies was the lack of a theory that would predict the effects that they should be looking for.

I have already mentioned that my interest in the geomagnetic effects began in Berlin in the summer of 1930. During my visit to the Reichsanstalt, Bothe, accompanied by Kohlhöster, went on an expedition to the North Sea and the northern Atlantic Ocean in the hope of discovering a variation of the cosmic-ray intensity with geomagnetic latitude. The prediction of such an effect was based simply on an argument of analogy with the phenomenon of the Northern Lights. These, at that time, were thought to be produced by streams of high-energy electrons originating from the Sun and channeled toward the polar regions of the Earth by the Earth's magnetic field. It was argued that, if primary cosmic rays were charged particles, the geomagnetic fields should exert upon them a similar focusing action.

Measurements taken between the geomagnetic latitudes of 51° and 81° gave a negative result. However, if the account of these measurements did not supply any new information, it was beneficial for me as it provided a motivation for trying to understand what actually should happen when an initially isotropic stream of charged particles enters the magnetic field of the Earth.

Bothe had called my attention to the very extensive work of Carl Störmer and his pupils who had worked for years on the mathematical problem of the motion of charged particles in the field of a magnetic dipole (which closely represents the Earth's magnetic field). But, as Bothe and Kohlhöster had mentioned in their paper, Störmer's theory appeared to be so complex as to rule out the possibility of applying it to cosmic-ray problems.

However, a careful examination of Störmer's papers convinced me that this was not at all the case, and that, by just asking the proper questions, it was easy to obtain from Störmer's theory some simple and highly significant results. Störmer's group, through years of painstaking numerical calculations (electronic computers, of course, were undreamed of, and Vannevar Bush at MIT was still in the process of perfecting his mechanical differential analyser) had computed the trajectories in the Earth's magnetic fields of hundreds of electrons of different energies originating from the Sun; their purpose was to find where and in which direction these electrons would strike the Earth's surface. But, cosmic-ray physicists were interested in a simpler problem; they wanted to know whether particles of a given energy could or could not reach a given point of the Earth in a given direction.

I found that the answer to this problem was contained, in essence, in a formula derived by Störmer which, for each point of the Earth and for each particle energy defines a cone with the axis perpendicular to the geomagnetic meridian. Known today as the *Störmer cone* (see Fig. 1.8), it separates the directions of arrival of the particles whose trajectories, followed backward, go to infinity, from the directions of arrival of the particles whose trajectories always remain in the proximity of the Earth. Clearly, the latter are forbidden directions for cosmic-ray particles, whereas the former

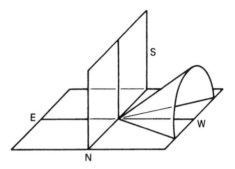

Fig. 1.8. The *Störmer Cone*, which separates the *allowed* from the *forbidden* directions. The axis of the cone is perpendicular to the geomagnetic meridian; the angle of its surface with the meridian plane is given by the equation:

$$\sin \vartheta = \frac{k^2}{R} \cos \lambda - \frac{2k}{R \cos \lambda}$$

In his equation, λ is the geometric latitude, R (= 6.38×10^8 cm) is the radius of the Earth; and, for particle velocities near the velocity of light, $k^2 = 300M/E^2$ where M is the magnetic moment of the earth (8×10^{25} gauss/cm) and E is the particle energy in electron volts. Forbidden directions are to the East of the cone for positive particles, to the West for negative particles.

are allowed unless the corresponding trajectories happen to intersect the Earth. The forbidden directions are to the east or west of the Störmer cone depending on whether the particles are positively or negatively charged.

On the basis of this result, which I published in *Physical Review Letters* in the summer of 1930, I predicted that the angular distribution of cosmic rays would show an asymmetry with respect to the plane of the geomagnetic meridian, with an excess of particles coming from the east or from the west depending on whether they carried a negative or positive charge, an effect that became known as the East–West effect. I also estimated the dependence of this effect on the energy of the particles and on the geomagnetic latitude of the place of observation.

Two questions were left unanswered. The first concerned the relative intensity of cosmic rays coming from different allowed directions and the second concerned the trajectories which, although coming from infinity, corresponded to forbidden directions because they intercepted the Earth.

The answer to the first question was provided by Fermi, who pointed out that, according to the Fourier theorem, if cosmic rays at large distances from the Earth had the same intensity in all directions (as was thought to be the case) then they would have the same intensity in all allowed directions at the Earth's surface.

The second problem was not amenable to an equally simple solution. Actually the question of the *shadow effect* of the Earth did not lend itself to an analytical treatment, but required an elaborate program of numerical computations. This was carried out in 1933, by George Lemaitre and Manuel Vallarta, using Bush's mechanical differential analyser. This program achieved its goal. But it did not materially affect the interpretation of the experimental data because the limited accuracy of the observations did not permit a verification of the details of the theoretical results.

Upon my return to Arcetri, in the autumn of 1930, I attempted to observe the predicted East–West asymmetry, but did not have any success. I was not greatly surprised because I knew that the asymmetry would become pronounced only at low geomagnetic latitudes and at a sufficiently high elevation to afford the observation of cosmic rays of comparatively small energy. Thus I started planning to repeat the experiment in the proximity of Asmara, a small town which was at that time the capital of the Italian colony of Eritrea.* This town rises on the high plateau of East Africa, at an altitude of 2370 meters and at a geomagnetic latitude of 11° 30' N. The project was enthusiastically supported by Garbasso. Sergio De Benedetti, then a young student, offered his co-operation, so that in a comparatively short time we were ready to start the experiment. Unfortunately, however, logistic problems delayed our departure to Africa. We had not yet left Italy when, in the fall of 1932, having won a competition for an academic position in an Italian university I was called as a professor of experimental physics to the University of Padua.

I left Arcetri with a heavy heart. I was young, and I knew that, in my life, there would be other periods of productive work and of rich human experiences. But I also knew that none of them would have that special flavor of my years in the Florentine hills.

* A northern province of Ethiopia conquered by Italy at the end of the last century.

2

Padua, Copenhagen, Manchester

Padua was regarded as one of the most prestigious cultural centers of Italy, and its reputation was deserved. The University of Padua had a proud history, stretching back to the early thirteenth century. It had been founded by a group of students and senior scholars of the University of Bologna who were hoping to find in Padua, then a free commune, those liberties (academic and personal) which their University, dominated by the church, could not provide.

The University of Padua soon flourished, attracting scholars from all parts of Italy and from abroad. When, some time later, Padua came under the rule of Venice, it turned out that the loss of independence by the city had a beneficial effect on the University. For now the University of Padua became the University of the powerful Venetian republic. As such, it was granted special privileges and, most importantly, it was afforded an effective protection from the intrusion of the church.

Through the centuries, the University kept faith to the liberal principles which inspired its founders. The student body always played an important role in the conduct of University affairs, such as the appointment of faculty members and the specification of the subjects to be covered in the various courses. The University of Padua was among the first (probably *the* first) to grant a doctoral degree to a woman, and to admit non-Christian students. Eminent teachers and distinguished students contributed to establish the fame of the University. Galileo taught there, undisturbed, for sixteen years (and if he had not left Padua he would have avoided the serious troubles which his theories brought upon him as soon as he was no longer under the protection of the Venetian republic); Niccolo Copernico had been a student in Padua as had future *dogi* (rulers of the republic) and many other prominent citizens.

Now the University had an excellent, not large, but well-balanced faculty, which covered all branches of knowledge, from sciences to letters, from medicine to jurisprudence. Several faculty members were internationally known scholars. Many were highly educated persons, whose interest and knowledge extended beyond the boundaries of their specialization.

The *Caffè Pedrocchi*, across the street from the sixteenth-century palace which housed the University, was a meeting place where one had the opportunity to participate in discussions that were enlightening and stimulating. Because of the small size of the faculty, everyone knew everyone else in the academic community, a circumstance which favored the formation of personal bonds of friendship.

Of my activities as a teacher, the one which gave me greatest satisfaction was my interaction with the advanced physics students. Some of them stood out for their personal and intellectual qualities. I remember Ettore Pancini, in whom I saw the promise of a future distinguished scientist. What I had not guessed was the depth of his political commitment and his disregard of personal dangers, qualities which became manifest during the war when he took a leading position in the armed resistance movement. I remember Giampiero Puppi. I had known him as a child in Venice and now I could clearly perceive the brilliant career that lay ahead for him. And I remember Eugenio Curiel, an intelligent, serious young man, one of our best students. I had offered him the opportunity to join the expedition to East Africa where I was planning to detect the East–West effect. He had accepted willingly; but a short time later I received a letter from him saying that he was no longer interested in physics, that he had decided to stop working for a doctorate in this field and that he did not wish to take part in the expedition to Africa. After a while, however, he came to see me to say that he had changed his mind and wanted to try getting a degree, which he soon did *summa cum laude*. This, indeed, was strange behavior on the part of a person whom I knew to be well balanced and of sound judgment. I was puzzled, and the puzzle was solved only some years later, when I learned that, at that time, Curiel was deeply involved in the clandestine antifascist movement and felt that he should devote his undivided attention to this cause. I have reason to believe that his

Consideriamo (fig. 65) in seno al conduttore un'areola elementare $d\sigma$ comunque inclinata rispetto al vettore densità di corrente $\vec{j}$; vogliamo calcolare l'intensità di corrente attraverso $d\sigma$. Consideriamo perciò un cilindretto avente come direttrice il contorno di $d\sigma$ e le generatrici parallele a $\vec{j}$; la sua sezione retta è $d\sigma \cos \vartheta$, se ϑ è l'angolo che la normale a $d\sigma$ forma con $\vec{j}$; e l'intensità di corrente in questo cilindretto, per la definizione stessa di j (formula (14)) è eguale a $j\, d\sigma \cos \vartheta$. Questa

fig. 65

rappresenta evidentemente anche l'intensità di corrente attraverso $d\sigma$, che è una sezione del cilindretto ; detta di tale intensità, si ha quindi in generale

(15) $$di = j\, d\sigma \cos \vartheta$$

o anche, indicando con $\vec{n}$ un vettore unitario normale a $d\sigma$:

(15') $$di = \vec{j} \times \vec{n} \cdot d\sigma.$$

La (14) può considerarsi come un caso particolare della (15), che si verifica quando $\vec{j}$ è perpendicolare a $d\sigma$.

Si abbia ora una superficie finita qualsiasi : proponiamoci di calcolare l'intensità di corrente attraverso ad essa. Basterà perciò suddividerla in tante areole elementari, calcolare mediante la (15) o la (15') l'intensità di corrente in ciascuna di queste areole, e sommare i singoli contributi. Otteniamo così, per l'intensità totale di corrente attraverso la superficie σ, l'espressione :

(16) $$i = \int_\sigma \vec{j} \times \vec{n} \cdot d\sigma.$$

Possiamo quindi dire che : " l'intensità di corrente attraverso una superficie σ qualsiasi è eguale al _flusso_ del vettore densità di corrente attraverso la superficie stessa,,.

I vettori, che rappresentano la densità di corrente nei vari punti di un conduttore, costituiscono un _campo vettoriale_. Analogamente a quanto abbiamo fatto per la forza elettrica, possiamo definire le _linee di corrente_ co

Fig. 2.1. A page of the lecture notes on electricity. Copied by hand and reproduced in lithography by CEDAM publishers.

brief return to academic work was due to the realization that he could not justify to his family his decision to renounce a degree, which was to be the crowning of a four-year effort, without disclosing the reason for this decision. Now the name of Curiel belongs to the list of the martyrs of the Resistance. Having fought during the war alongside the partisans, in the capacity of a commander, he fell in Milan, a victim of the Germans, on the eve of the liberation.

Objectively, I knew that I had reasons to be grateful to the University of Padua for the friendly welcome by the faculty and for the many opportunities which I was being offered. And yet, when I think of my past, I remember my years in Padua as a 'grey' period of my life. Perhaps I was missing the 'spirit of Arcetri'. A more tangible factor was the slackening of my research activity which I experienced in the passage from Florence to Padua. Such a slackening was clearly a consequence of the change of my academic status, for now, as a professor, I had many duties which prevented me from devoting all my time to research, as I had done when I was just an assistant.

This does not mean that none of my new duties had any gratifying aspects. As a matter of fact, teaching, the most demanding of these duties, held a great interest for me. I devoted with pleasure a sizeable part of my time to it. I would carefully prepare my lectures and would deliver them as well as I could. I would write the lecture notes which were then copied by hand and reproduced in lithography (Fig. 2.1).

Another task which I undertook willingly, although not as enthusiastically as teaching, was the planning of a new Physics Institute and the supervision of its construction. When I arrived in Padua, the Physics Institute was housed in a few rooms of the ancient University palace. To this location were attached important historical memories. Among other things, a chair was kept there from which it was believed that Galileo had done his teaching. But, by now, these premises had become utterly inadequate to the needs of physics both because they did not provide sufficient space, and because they did not allow the installation of modern equipment. Therefore, the University had decided to build a new Institute, and the rector had asked me to take on the responsibility for this project.

I had accepted this assignment determined to see to it that the new Institute would be what, in my mind, a modern Physics Institute should be.

I knew that I was faced with a difficult problem. Thus, I went to visit several Italian and foreign institutes, to see how others had dealt with problems similar to mine. Then I sat down with an architect, Daniele Calabi, and began making detailed plans for the Institute. We carefully studied how to organize the facilities, where to locate the various rooms in view of their purposes, how to avoid superfluous spaces, etc.

The Institute was inaugurated in 1937 and won the praise of whoever saw it. (Perhaps Franco Rasetti from the University of Rome remembers when, having come for a visit, exclaimed: 'This is indeed a physics institute, not an *Opera del Regime*! [a work of the fascist regime]' But the story of the Institute has a sad ending, for, a short time after its opening, its doors were shut to me by the anti-Semitic laws. Others would use my Institute, but, as far as I was concerned, the time and effort I had spent in creating it had been wasted.

Although my research activity lost some of its momentum when I moved from Florence to Padua, it by no means came to a stop. My first concern was to carry out the experiment on the East–West effect, which had been delayed by my move to Padua. As I have already said, the instruments for this experiment had been prepared while I was still in Arcetri. After a last check, I had them shipped to Venice where, taking them along, I boarded a freighter which was about to sail south. I was accompanied by Sergio De Benedetti, who had followed me from Florence. After about a week, we landed in Massaua, on the Red Sea, in Somaliland. From there a motor vehicle of the Italian army took us to Asmara.

At those times, Eritrea appeared to be a quiet land, relatively well-off. The features of the natives (of Camitic stock), were reminiscent of those of the populations of southern Europe. Many of the natives worked for the Italian colonists in the coffee plantations; others had been enlisted by the army in a special corps called the corps of the *ascari*. It was not unusual to find, in the local population, persons with some degree of education. All in all, there were good relations between the Italians and the natives. I did not see any sign of the tension, which, I was told, existed in other colonies where racial prejudice relegated the natives to a condition of humiliating serfdom.

Later, I had a chance to go further inland. Here, the country was much wilder, and the population much more primitive; the physical features of the natives were characteristic of the Negro race. Also in this region, however, the healthy aspect and the peaceful disposition of the population bore witness to some degree of well being.

For our experiment the army engineers had built a wooden cabin, on top of an *amba* (a flat-top hill, similar to the mesas of the American southwest) a short distance from Asmara. A tent was installed to protect the cabin from the equatorial sun. A temporary line supplied electric power.

Our experiment consisted of measuring the intensity of cosmic rays in different directions. The instrument used for these measurements was a simple 'cosmic-ray telescope', formed by two counters operated in coincidence, and placed horizontally at a suitable distance from one another (Plate 22). The counters were attached to a rigid steel frame, mounted on pivots, which made it possible to point the axis of the 'telescope' in any desired direction.

In some of the measurements the counters were enclosed in heavy lead sleeves which, by shutting off the soft component of cosmic rays, were expected to enhance the predicted asymmetry. I argued that only the hard component (at that time believed to be of primary origin) should exhibit this asymmetry, while the soft component (known to be of secondary origin) was expected to be uniformly distributed in azimuth, so that its presence would tend to obscure the asymmetry. As we shall see, the hard component is also of secondary origin. That it exhibits the East–West asymmetry predicted for the primary radiation is explained by the fact that the hard secondary particles are ejected in directions close to the direction of the primary particles by which they are produced.

As soon as we started the observations it was clear that the cosmic-ray intensity was greater in the western than in the eastern direction. This asymmetry unequivocally confirmed the corpuscular theory of cosmic rays for it showed that the primary radiation consists, at least for the most part, of electrically-charged particles. But at the same time, the results were surprising because those of us who had supported the corpuscular theory were convinced – more, I must admit, from prejudice than because of a logical argument – that the primary particles would turn out to be electrons, and would therefore be negatively rather than positively charged. We continued the observations for several weeks, alternating measurements with and without the lead sleeves, and with the axis of the 'telescope' pointing in different directions.

These measurements confirmed the predicted enhancement of the East–West effect by the lead sleeves. The most pronounced asymmetry was observed with the counters inside the sleeves and with the axis of the 'telescope' at 45° to the vertical. In these conditions we found that the cosmic-ray intensity to the west of the geomagnetic meridian exceeded that to the east by about 20 percent.

Here I must digress to report that, when we were about to leave for Eritrea, there appeared in *Physical Review* two letters to the Editor, one by Thomas Johnson and one by Luis Alvarez and Arthur Compton relating the observation of an East–West effect at Mexico City. So, for a few months, we had lost the priority of this important discovery. Moreover, the authors had ignored my prediction of this effect and had given credit for it to Lamaitre and Vallarta, whose paper had been published three years after mine. I took the view that the overlooking of my work was to be explained as an oversight. But if this interpretation (confirmed several years later by an apologetic letter from Alvarez) prevented my resentment, it did not alleviate my disappointment.

A further result of our observations in Eritrea deserves to be mentioned. I translate from the original paper:

> The frequency of the coincidences recorded with the counters at a distance from one another, shown in the tables as 'chance coincidences' appears to be greater than would have been predicted on the basis of the resolving power of the coincidence circuit. Those observations made us question whether all of these coincidences were actually chance coincidences. This hypothesis appears to be supported by the following observations . . . Since the interference of possible disturbances was ruled out by suitable tests, it seems that once in a while the recording equipment is struck by *very extensive showers* of

particles, which cause coincidences between counters, even placed at large distances from one another. Unfortunately, I did not have the time to study this phenomenon more closely.

This, I believe, was the first observation of those extensive air showers which, a few years later, were studied in some detail by Pierre Auger and his collaborators and which, more recently, became the object of a major research project by the MIT cosmic-ray group.

Another problem which commanded my attention during my first years in Padua was the mechanism of shower production. Many experimental data, obtained to a large extent in Arcetri, were available. Needed now was an in-depth analysis of these data, which would allow us to reach reliable conclusions about the nature of this phenomenon.

I applied myself to this analysis. The most significant result, which I published in 1934, may be stated as follows. In the local cosmic radiation there exist two separate components, entirely different in their nature and properties. The first – the so-called *hard component* – consists of electrically-charged particles, which lose energy in matter exclusively through ionization processes and which, therefore, have a range practically proportional to their energy. The second component consists of rays whose interaction with matter manifests itself primarily through the production of showers. Their ability to produce showers shows that the individual energy of the rays is very large. The fact that, in spite of this, their range is very small, is due clearly to the absorption which they undergo by the production of showers.

In those years, as I have said repeatedly, it was still believed that the hard component consisted of particles of the primary cosmic radiation which had penetrated the atmosphere. There remained the problem of finding the nature and origin of the shower-producing radiation. At first we thought that this radiation was of secondary nature, being produced by the hard components. De Benedetti and I undertook some preliminary experiments which cast serious doubts on this hypothesis. To settle the question once and for all, we decided to compare the variation with altitude of the frequency of showers with the variation with altitude of the intensity of the hard component. In the summer of 1934 we carried out measurements in Padua (40 meters above sea level), and in the Alps, at *La Mendola* (1350 meters) and at *Lo Stelvio* (2700 meters).

We found that the shower frequency increases with altitude much more rapidly than the intensity of the hard component (from Padua to *Lo Stelvio* an increase of about a factor of three for the showers, against a factor of about 1.5 for the hard component). This result showed that the shower-producing radiation was not a secondary radiation of the hard component. In the absence of other ideas, we were led to conclude that the shower-producing radiation must be part of the primary radiation.

Beginning in 1934 there appeared in rapid succession a number of theoretical and experimental results which substantially changed our views about cosmic-ray phenomena. The theory developed by the German physicists Hans Bethe and Walter Heitler showed that the shower-producing radiation consists of high-energy electrons and photons, and that the showers themselves are produced by alternating processes of photon production by electrons and materialization of photons, giving rise to pairs of negative and positive electrons. In addition, the same theory showed that the particles of the hard component were 'new', previously unknown particles with a mass greater than that of electrons, but smaller than that of protons. These particles of intermediate mass became known as *mesotrons*. These conclusions were soon experimentally confirmed by cloud-chamber observations. With this technique the Harvard physicists J. C. Street and G. C. Stevenson found that the mesotron mass was about 175 times the electron mass (we know today that it amounts to 206.8 electron masses).

I followed these developments with great interest, if with some regret for not having been in a position to contribute actively to them. Several circumstances hampered my scientific activity after the mid-1930s. Foremost among them was a growing concern over political events, which prevented me from focusing my mind on problems of physics.

The sky over Europe was dark and becoming darker day by day. But, in the midst of the pervading gloom, my own horizon was growing brighter. I saw Nora for the first time in Venice, at the wedding of one of her cousins. She claims that she did not take any notice of me in the crowd. But I know that I did notice her. Some time later, in 1937, we chanced to meet at the Lido, where she was spending her summer vacations. We came into the habit of getting together on the beach, alone or in the company of some common friends. What happened next is a timeless story. Increasingly we found pleasure in each other's company. Increasingly we enjoyed doing things together, just being together. We were married the following April and, through rain and shine, this marriage has survived until this day, for over half a century.

At first trying to close our eyes to what was going on around us, we busied ourselves making plans for what we thought would be our life in Padua. We rented an apartment, we furnished it, we began to make contacts with the local community. But as summer wore on the threats of the developing situation could no longer be ignored.

In Europe, the likelihood of a general conflict was growing increasingly alarming. In Italy a policy of discrimination against the Jews was rapidly gaining momentum. This was a preposterous and entirely unexpected development. In modern Italy, anti-Semitism had been unknown. The Jewish community, although rather influential, was quite small (about one-thousandth of the whole population). Whether still faithful to the ancient religion, or converted to Christianity, or behaving as agnostics, Jews were fully integrated into Italian society.

Jews had fought, alongside other Italians, in the wars for the independence of Italy. In the unified nation born of these wars, Jews had held highly responsible positions in the government and had partaken effectively in the artistic, literary, scholarly, and economic life of the country. The world-renowned Italian school of mathematics was, to a large extent, the creation of Jewish scholars.

This deep-rooted tradition of brotherhood between Jews and non-Jews in Italy held fast until the middle 1930s when, under the coercion of Nazi Germany, the first signs of a burgeoning anti-Semitism began to appear. Few people expected that these signs should portend an official anti-Jewish policy of the fascist regime, a policy which would have been greatly unpopular. But public opinion was wrong. Soon decrees were enacted which, one by one, deprived the Jews of their rights as Italian citizens. Eventually, in September of 1938, I learned that I no longer was a citizen of my country, and that, in Italy, my activity as a teacher and as a scientist had come to an end.

To describe this event as a tragedy would be a gross exaggeration; at that time truly tragic was the fate of so many people in Europe. But it was a hard blow. Fortunately I had friends abroad, and I was confident that, with their help, I would be able to start a new life in some foreign country. It was not easy for either of us to leave Italy without knowing whether we would ever be able to return, without knowing what would happen to our families that stayed behind. But Nora, with a clear perception of the serious danger, which loomed in the future if we stayed in Italy, insisted that we leave as soon as possible.

The trouble was that neither one of us had a passport and at that time (we were in the midst of the Munich crisis) to obtain a passport was very difficult for anyone and practically impossible for a Jew. But in Italy (as I was to learn once more) in case of need one will always find someone who is willing to help and knows how to do it. I remembered that a few years before I had received from the Royal Italian Academy a fellowship for a period of study abroad. With this fellowship I had gone to Paris, where I had worked in the private Institute that the Duc Maurice de Broglie had created in his palace. (Incidentally my friendship with the French physicists Louis Leprince-Ringuet and Pierre Auger dates to those few months in France.) I also remembered that I had cut my stay abroad short because of the call to the University of Padua. Thus I wrote a letter to professor Giancarlo Vallauri who had succeeded Marconi as the president of the Academy, saying that I would like to use the remainder of my fellowship for a second period of study abroad. I also pointed out that, to go abroad, I needed a passport.

After a few days I received a letter from the Chancellor of the Academy which read: 'On behalf of His Excellency Vallauri I inform you that he authorizes you to use the last installment of the fellowship that was granted to you some time ago in order to perform the mission suggested in your letter.' A short time later came the passport and a sum of money. Nora could not use the same pretext to obtain a passport. But she managed to do so by some other, typical Italian means, as she herself will explain. While we were trying to obtain our passports, I had written to Niels Bohr telling him that I would like to spend some time at his Institute in

Copenhagen. Bohr had answered immediately, saying that he would be happy to have me as his guest. And so, on October 12, we left for Copenhagen, the first station in our wanderings.

We remember Copenhagen as a pleasant, hospitable city. Above all, we gratefully remember the warm welcome we received from Niels and Margrethe Bohr, and from the people around them. The two months we spent in Copenhagen were a most beneficial interlude. The human interests, the lively intellectual climate, the sane view of political events that were the essence of the 'spirit of Copenhagen' went a long way toward clearing our minds and strengthening our confidence in the future. I spent long hours in the library bringing myself up to date on the recent developments in physics, talking with the people I happened to meet, thus, gradually rekindling my enthusiasm for science.

While we were in Copenhagen, Bohr called a meeting which was attended by a large number of scientists, among them many cosmic-ray physicists. I could swear that one of Bohr's motives was to give me the opportunity of meeting people who might help me find a job. In any case, this is exactly what happened, because shortly thereafter Patrick Blackett wrote, inviting me to Manchester with a fellowship of the *Society for the Protection of Science and Learning*.

Before I speak of the new phase of our peregrinations, which followed this invitation, I would like to relate an episode of our stay in Copenhagen. Bohr knew (or guessed) my financial straits. Apparently he wanted to help me, but did not dare to do so directly for fear of hurting my feelings. Thus one day, shortly before the conference he approached me in the library and asked me to go to his secretary. I obeyed, and she handed me an envelope with a sum of money. I was surprised and asked for an explanation. 'It is for your travel expenses to the conference', she answered. 'But I am already here, so I don't have any travel expenses' I said. 'True, but if you had not been here we would have had to reimburse your expenses.'

In mid-December, after a stormy crossing of the North Sea, we arrived in England and headed for our destination in Manchester. Manchester was then a squalid city, as different from Copenhagen as one could dream. Blackett was conscious of this, and was wondering whether we Italians could survive in such a place. And yet both Nora and I have fond memories of Manchester.

True, our living conditions were uncomfortable. True, the city was depressing, with a wintery sky permanently dark, with a moist atmosphere, with the omnipresent signs of the disintegration of a once flourishing textile industry. But there were compensating pluses. Contact with a simple and genuine population motivated by a feeling of human solidarity; the warm relationship with Patrick and Constance Blackett; the presence of other friends, like the Brentanos, in whose house we took shelter when the cold chased us from our apartment. Moreover, with the coming of spring, the moors around the city became a sea of flowering daffodils, and we spent our weekends exploring them by bicycle, with benefits to our souls and our bodies.

Blackett's laboratory was a very active research center. Working there were several of Blackett's former pupils, now themselves mature scientists, among them Bernard Lowell, George Rochester, and J. C. Wilson. The Hungarian physicist Ludwig Jánossy, who had left his country for the same reason as I had left mine, was also a guest of the laboratory. Blackett came often, and his presence was an incentive and guide in our work.

Very important for me was to have a chance of starting some experimental work again. Jánossy and I measured the absorption in lead of the gamma-ray component of cosmic rays. Of course these measurements did not produce any world-shaking results; they only confirmed the theory of Bethe and Heitler. But ours was a clean experiment, elegant in its simplicity. Moreover it gave us the occasion to develop a new technique, which we named the *anti-coincidence technique* and which was to find many applications in later experiments.

Here, perhaps some explanation is called for. The basic problem was that of separating the gamma-ray component from the much more abundant corpuscular component of the cosmic radiation. The experimental arrangement used for this purpose is shown in Fig. 2.2. Gamma-rays which traverse the absorber (S) will go through counters A undetected because they do not ionize the gas in these counters. Some of the gamma-rays will then undergo materialization in the lead screen (s), producing pairs of electrons. These, in turn, will discharge counters B, C and D. The charged

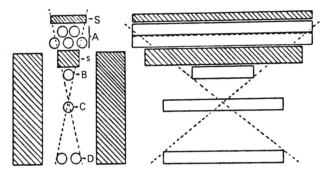

Fig. 2.2. An early application of the anti-coincidence method. A simultaneous discharge of counters *BCD*, which is *not* accompanied by a discharge of counters *A* (anti-coincidence *BCD–A*) identifies a gamma-ray photon which after traversing unrecorded counters *A* produces in the lead screens a pair of electrons that discharge counters *B*, *C*, and *D*. From a paper by L. Jánossy & B. Rossi in *Proceedings of the Royal Society*, A960, 175–88 (1942).

particles of the corpuscular radiation, on their part, will also discharge counters *B*, *C* and *D*, but in addition will also discharge counters *A*.

Therefore one could find the number of events produced by gamma-rays by subtracting the number of coincidences *A B C D* (produced *only* by the corpuscular component) from the number of coincidences *B C D* (whether or not accompanied by a discharge of counters *A*) which are produced by *both* the corpuscular *and* the gamma components.

However, this procedure, which calls for measuring a small quantity (here the gamma-ray flux) as the difference between two large quantities, is not apt to produce accurate results. It is suggested by intuition (and is proven by the mathematical theory of statistical errors) that one can achieve a much higher degree of accuracy by *directly* measuring the gamma-ray flux. For this purpose we developed a special *anti-coincidence circuit* so designed as to detect the events *B C D* – A, i.e., the coincidences *B C D* that are not accompanied by a simultaneous discharge of counters *A*, which can be produced only by gamma-ray photons. The experiment then, consisted of constructing the absorption curve of gamma-rays by plotting the rate of anti-coincidences *B C D*

– A against the thickness of the absorbers (S) placed above counters A. As I mentioned, the results were in agreement with the theoretical predictions.

In the meantime a new and interesting problem in the physics of elementary particles had come to the fore. In 1935, the renowned Japanese scientist Hideki Yukawa had developed a theory of nuclear forces which predicted the existence of particles with mass intermediate between the masses of the electron and the proton. In order to explain the β decay, he had postulated that these particles should be unstable, with a mean life on the order of microseconds, and that their decay should each give rise to an electron and a neutrino.

Yukawa's theory, associating a new particle with the nuclear field of force, was a brilliant concept, of fundamental importance for future developments of theoretical physics. It was natural to identify the newly discovered mesotron with Yukawa's particle. However it turned out that this identification was incorrect, as were the decay scheme and the mean life predicted by Yukawa for his particle. But two incorrect assumptions led to a prediction which, as we shall see, turned out to be correct; namely that mesotrons are unstable particles, with a mean life on the order of microseconds.

In 1939, however, this prediction had not yet been verified experimentally, and was one of the hottest problems for particle physicists. There had been attempts to observe the electrons that were supposed to be emitted in the decay of mesotrons, but these attempts had not been successful. On the other hand, it looked as if the assumption of mesotron instability might receive support from some experiments which suggested the occurrence of an *anomalous* absorption of mesotrons in the atmosphere; meaning that mesotrons appeared to undergo a stronger absorption in air than in condensed materials. Such anomalous absorption was to be expected if mesotrons were unstable particles with lifetimes of microseconds; for then many mesotrons, travelling through air, would decay before being stopped by ordinary energy losses by ionization, whereas no mesotron would have the time to decay before reaching the end of its range in condensed matter.

Blackett and I carefully examined the experimental results that were quoted in support of the evidence for an anomalous absorption of mesotrons (variation of the anomalous absorption of mesotrons with height, with zenith angle, with atmospheric temperature). We reached the conclusion that none of these results (with the possible exception of some questionable data concerning the zenith angle variation) could be regarded as proof of mesotron instability.

The British period of our lives as exiles was coming to an end. For some time I had been in correspondence with Arthur Compton. He urged me to go to Chicago to participate in a symposium on cosmic rays planned for the next summer. He also told me that he was trying to find a job for me, but this was a difficult task because of the large number of scholars converging from Europe on the United States. We were sorry to leave Europe. We were especially sorry to leave England, where we had found such warm hospitality. I was also concerned about our going to America without knowing how I would be able to make a living. But the international situation was growing more and more tense. Blackett was very worried for us and strongly advised us to leave. And so, in the middle of June 1939, we boarded the French liner 'Liberté' bound for America.

We stopped for a couple of weeks in New York. Enrico and Laura Fermi had arrived there a few months before us, and we felt the need to reestablish contact with some old friends in order to overcome the shock of the transition from one continent to another.

In those days Hans Bethe was also in New York. He, too, was an old friend. I had known him since the times when, on his way to Rome to visit Fermi, he would stop in Florence to talk with us in Arcetri . . . and to learn to appreciate Italian cuisine at my mother's home. We were happy to see him again. He was about to drive to Chicago, and offered us a passage. We accepted with great pleasure.

3

Physics of elementary particles in the Age of Innocence

(1939–1943)

We arrived in Chicago just in time for the opening of the cosmic-ray symposium that Compton had organized. Many European scientists had come to Chicago for the occasion, Bothe among them. I was glad to see him again, but, unavoidably, political events had cast a shadow over our personal relationship. I remember that Bothe was disturbed because I no longer would communicate with him in his tongue, as I always used to do in the past. Probably he took this as a sign of hostility toward Germany. (Bothe, while not sympathizing with the Nazi regime, had nonetheless a strong sense of loyalty toward his country.) Actually what had happened was that the effort of learning English had entirely erased German from my memory.

A whole day of the symposium was devoted to the problem of the radioactive instability of mesotrons. Many papers were presented, many experimental data were discussed in detail. The consensus was that the evidence for this phenomenon could not yet be regarded as decisive. Therefore further and better experiments were needed, and the first in line was a positive verification of the predicted anomalous absorption of mesotrons in the atmosphere.

I thought that the most promising approach was an accurate and direct comparison between the absorption of mesotrons in air (where, if mesotrons were unstable, both energy losses by ordinary ionization processes, and disappearances by decay in flight would contribute to the total attenuation) and the absorption of mesotrons in condensed matter (where mesotrons would not have time to decay before reaching the end of their range and where, therefore, the attenuation would be due to ionization losses alone). Clearly, the most critical aspect of the experimental program was a precise estimate of the absorption of mesotrons in air; and it appeared that the only method to achieve this purpose was a measurement, in the vertical direction, of the flux of mesotrons at various atmospheric' depths, i.e., at various altitudes above sea level.

After the symposium, Arthur and Betty Compton had kindly invited us to spend a few days in their summer cottage at Otsego Lake in upper Michigan, so that we might recuperate from the stresses and the emotions of the last weeks. I took this occasion to tell Compton of my thoughts about the supposed instability of mesotrons, and of my belief that mountain experiments would offer the best opportunity for reaching a final decision about the occurrence of this phenomenon.

Compton listened with interest. He then pointed out that the region of the Rocky Mountains of Colorado was the ideal place for an experiment such as I had in mind. There was a road reaching the top of Mt Evans, over 4000 meters above sea level. At the top, a cabin had been built a few years before for the use of the scientists working there. We could count on the support of Joyce Stearns, professor of Physics at the University of Denver.

Aroused by what I had heard from Compton, I asked him if he thought that it might be possible to organize an expedition to Colorado the next summer. His answer was: 'why not *this* summer?' I was taken aback. It was almost the middle of July, and I had previously agreed to spend a week at the summer school in Ann Arbor after leaving Otsego Lake. On the other hand, I would have to be on my way before the end of August in order to have a chance to complete the experiment before the first snows blocked the mountain roads. Thus I had merely one month for building the experimental equipment and making the necessary logistic arrangements. Compton's well-meaning suggestion had placed me in a difficult spot. But, of course, I did not draw back; and, immediately after my return to Chicago, I began working with a sense of urgency.

Compton had asked two physicist friends of his, Norman Hilberry of New York University and Barton Hoag of the University of Chicago, to help me in the preparation and execution of the experiment. We built our G.M. counters, then I started wiring a coincidence circuit on a breadboard as I used to do in Italy. Hilberry looked with undisguised disgust at the crude instrumentation which

I was putting together. 'Here in America' he said 'we build our circuits neatly on metal chassis, which we then mount on special racks.'

At first I was overwhelmed. How could we possibly build an elaborate instrument, such as Hilberry had in mind, in the very short time at our disposal? And without my having ever made anything of that kind? But soon I relaxed and decided that I would continue to work at my breadboard circuit, while letting Hilberry build his American-style instrumentation. It is no reflection on Hilberry's competence but rather on his perhaps excessive optimism that his ambitious project could not be completed in time, so that we had to leave for Colorado with my crude device.

Transportation from Chicago to Colorado was not a trivial problem. In 1939, we were still at the tail end of the Great Depression and very little money was available. Thus to buy or rent a truck was out of the question. Therefore Compton arranged with the Zoology department to borrow an old bus used during the school months for taking students on field trips, and in this bus we loaded our equipment.

Our experimental arrangement was very simple (see Fig. 3.1). It consisted of three G.M. counters placed horizontally one above the other and separated by a lead filter of sufficient thickness to absorb the soft component of the cosmic radiation. Lead shields on the side of the counters acted as effective screens against air showers. With this arrangement one could be sure that threefold coincidences could be produced only by particles of the 'hard' component, i.e., by mesotrons.

A carbon absorber, made of bricks of graphite, could be placed above the counters. Since carbon has an atomic number close to that of air, we argued that charged particles would lose practically the same amount of energy in ionization processes while traversing layers of carbon and air of the same mass per unit area.

The end of August was approaching, and we were ready to go. At the last moment Compton had decided that we would need the help of a strong young man to move our heavy equipment, and thus had asked a graduate student to accompany us. He was a member of the university swimming team and one of the best athletes on the campus. His name was Winston Bostick and his presence in our small group turned out to be particularly useful and pleasant. I was

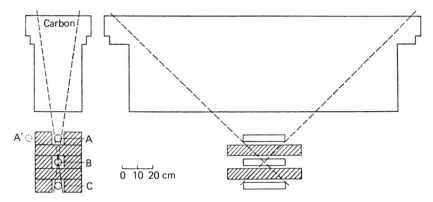

Fig. 3.1. Arrangement of counters and absorbers of the experiment which established the existence of an anomalous absorption of mesotrons in the atmosphere, thereby proving the radioactive instability of mesotrons. Coincidences between the G.M. counters A, B, C are produced by mesotrons which traverse the three counters and the interposed lead absorbers. From an article by B. Rossi, N. Hilberry & J. B. Hoag in *Physical Review*, **57**, 461 (1940).

happy to see him again recently, at a cosmic-ray meeting; he had made a name for himself with his work on the physics of plasma.

We left Chicago on August 26. One month earlier I had received from the University of Chicago the offer of a Research Associate-ship, with a yearly salary of $2500, paid by the Committee in Aid of Displaced Foreign Scholars. I had accepted this offer with a feeling that, by so doing, I had made a final decision to remain in the United States.

Nora and I, together with Hoag, travelled in the bus which, not being a particularly fast conveyance, could not make it to Denver in less than three days. The memory of that trip, through the limitless midwestern plains, among unending fields of corn and wheat – our first contact with the very heart of America – is still alive in our minds.

The travellers in the plains will not notice that the ground rises gradually, but steadily, toward the west. Denver, at 1600 meters altitude, still lies on perfectly flat ground. But only a few miles west of the city, abruptly rise the majestic Rocky Mountains, the power-ful wall that cuts across the American continent.

In Denver we were received, as if we were old friends, by Joyce Stearns and his family. We stopped there for a couple of days to take some measurements; then we reached the mountains and proceeded to Echo Lake, a small picturesque lake at 3200 meters above sea level, surrounded by a dense forest of spruce. There we stopped for a short time to take some more measurements and then, on September 1, we started for the top of Mt Evans.

The last leg of our trip was quite adventurous. In those days the paved road stopped at Echo Lake. From there to Mt Evans there was only a narrow, steep, rocky path, barely accessible to motor cars. Our bus, which, until then, had behaved quite well, was now beginning to show signs of discomfort. Nor could we blame it; after all it was not used to carrying tons of lead and graphite on such roads, up the highest mountains of the United States. As a precaution, before leaving Echo Lake, we filled all available containers with water. Shortly after beginning the ascent the water in the radiator started to boil. Nora took charge of the problem. Sitting on the hood, she began pouring water in the radiator, husbanding our water supply to make it last until we reached Summit Lake, a small lake above the timber line, about half way between Echo Lake and Mt Evans. There we got a fresh supply of water, which enabled us to reach the top safely. The only trouble was the noisy protest by the many cars behind us, which did not dare to pass our bus and were thus obliged to follow us at a snail's pace.

The summit of Mt Evans, at 4300 meters above sea level, is split into two peaks. Our cabin was on one of these. The road ended on the saddle between the two peaks, and there we parked our bus. In order to avoid lugging our heavy equipment up to the cabin, we decided to carry out our measurements in the bus, which thus became our itinerant laboratory. The cabin would be used as our living quarters.

It is well known that altitude has, at first, an exhilarating effect; it creates a sort of intoxication, and an almost irresistible craving for motion, for action. Nora and I, thanks to our excursions in the Alps, were well acquainted with the risks as well as with the charms of the mountains; and we also knew that one had to avoid any excessive effort until the body had become adjusted to the high elevation. But our companions had never seen a high mountain before and, without heeding our warnings, they started running up and down between the saddle and the peak, determined to carry all our supplies to the cabin immediately. The result was a violent attack of mountain sickness which put them out of commission for a full day.

For a couple of weeks we kept driving up and down the mountain, taking measurements alternately at Mt Evans and at Echo Lake. Around the middle of September we drove down to Denver and by the end of the month we were back in Chicago. At each station (Chicago, Denver, Echo Lake, and Mt Evans) we measured the rate of threefold coincidences between the G.M. counters, both with and without the graphite absorber above them. From the measurements taken at different altitudes without the graphite absorber it was possible to determine the attenuation of mesotrons in the atmosphere. The measurements taken at each individual station with and without the graphite absorber gave us the attenuation of mesotrons in carbon.

The results of our measurements are shown graphically in Fig. 3.2, where the quantity plotted on the horizontal axis represents the total mass in grams per square centimeter of air and carbon above the counters, and the quantity plotted on the vertical axis is the observed counting rate on a logarithmic scale. The open dots refer to measurements taken at the four stations without the graphite absorber. Thus the solid line connecting these dots represents the attenuation of mesotrons in the atmosphere. (Incidentally, no similar measurements of comparable accuracy had ever been performed before.) The solid dots refer to measurements taken at each of the three higher stations under the graphite absorber. Thus the dotted lines connecting the open and solid dots represent the initial slopes of the absorption curves of mesotrons in carbon. We see from the graph that mesotrons undergo a much stronger attenuation in air than in carbon when equal masses per square centimeter of the two substances are compared. We had thus achieved the first unambiguous demonstration of the anomalous absorption of mesotrons in the atmosphere, therefore proving their radioactive decay in flight.

On September 30, as soon as we had completed the analysis of our results, we sent a letter to the *Physical Review*, in which we wrote '*we see therefore definite evidence for the disintegration of mesotrons.*' The gratification which transpired from these few lines was, I believe, justified. This was the first time that the radioactive

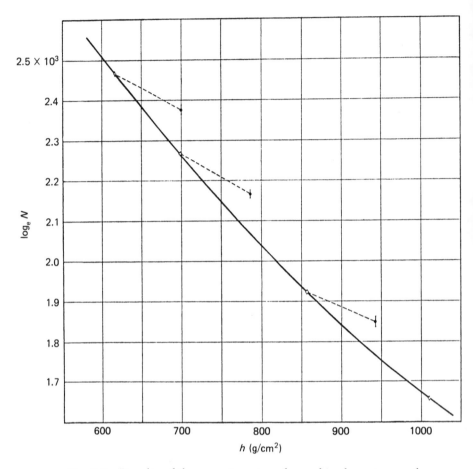

Fig. 3.2. Results of the experiment performed in the autumn of 1939 with the arrangement shown in Figure 3.1. The horizontal axis (the *abscissa*) shows the total mass in grams per square centimeter of air and graphite above the counters. On the vertical axis (the *ordinate*) is plotted, on a logarithmic scale, the flux of mesotrons as measured by the coincidence rate. The solid line connecting points representing results of the measurements taken at four different heights without the graphite absorber is the absorption curve of mesotrons in the atmosphere. The dotted lines, connecting the points representing measurements taken at the same height with and without the graphite absorber show the absorption of mesotrons in graphite at three different heights. One clearly sees that the absorption in the atmosphere is much greater than the absorption in graphite. The difference in the absorption of air and graphite is due to the decay of mesotrons in flight. From a paper by B. Rossi, N. Hilberry & J. B. Hoag in *Physical Review*, 57, 461 (1940).

instability of a subnuclear particle had been experimentally demonstrated.

Shortly after the publication of our results, I read the reports of several experiments, similar, in principle, to ours, which had been performed in the USA and in Europe. Among the latter were some measurements taken in the Alps by a group of old friends of mine. All of these experiments agreed qualitatively with ours in suggesting the existence of an anomalous absorption of mesotrons in air. Quantitatively, however, the results showed strong discrepancies, easily explained by the low accuracy of the measurements.

The radioactive instability of mesotrons, besides being a phenomenon of obvious importance for the physics of elementary particles, also provided the interesting opportunity of performing an experimental test of Einstein's relativity theory. A well-known prediction of this theory is the so-called dilation of time intervals, a phenomenon whereby the rate of advance of a clock in motion should appear retarded, as seen by a stationary observer. Until then it had not been possible to verify this prediction experimentally because of the lack of 'clocks' moving at a sufficiently high speed. Now, with the unstable mesotrons, nature had provided us with 'clocks' flying by at a speed close to that of light; thus an experimental test of the relativity prediction had become feasible. I decided to take advantage of this opportunity by starting a research program to study the velocity dependence of the mean life of mesotrons before decay. Since the details of this program cannot be presented in simple words, it may be helpful if I begin by spelling out the results that were obtained.

The first result, of a qualitative nature, was an experimental proof that the mean life of mesotrons before decay does actually increase with increasing velocity, as predicted. The second result was a quantitative verification of the relativistic equation describing the dependence of the mean life on velocity. As a by-product, the experiments provided an approximate estimate of the mean life of mesotrons at rest, which turned out to be on the order of a few microseconds.

What follows is a brief account of the procedure which produced the results mentioned above. The starting point was an equation, derived from the theory of relativity, which gives the mean life before decay of mesotrons in motion (τ) as a function of their mean life at rest (τ_0), and of their velocity (v). This equation reads

$$\tau = \frac{\tau_0}{\sqrt{1 - v^2/c^2}}$$

(3.1)

which shows that τ increases with increasing v, but becomes appreciably larger than τ_0 only when v approaches the velocity of light (shown as c in the equation). This equation, as such, cannot be tested experimentally for neither the mean life (τ), nor the velocity (v) can be measured directly. Clearly, we must base the experimental proof of the theoretical predictions on an equation which expresses these predictions in the form of a relation between observable quantities.

To obtain this equation, we multiply both sides of eqn (3.1) by the velocity (v). We note that the product (τv), that I shall denote by l, represents the mean free path before decay of the moving mesotrons, i.e. the average distances they travel before disappearing by decay. We also recall that the momentum (p) of a moving particle of mass m and velocity v is given by the equation.

$$p = \frac{m v}{\sqrt{1 - v^2/c^2}}$$

Therefore eqn (3.1) may be rewritten in the simple form

$$l = \frac{\tau_0}{m} p$$

(3.2)

We can then verify the prediction of the relativity theory concerning the dilation of time intervals by proving the proportional relation between l and p expressed by eqn (3.2). This proof is experimentally possible because l, the mean free path before decay, can be deduced immediately from a measurement of the anomalous absorption of mesotrons in air (which is due to the decay of mesotrons in flight) while momentum (p), can be obtained from a measurement of the range of mesotrons in matter, by means of the well-established theory giving the energy loss of charged particles by ionization.

With these results in mind, during the summer of 1940 I went back to Colorado accompanied by two young physicists, David Hall and his wife Jane. Nora was also with us but, since she was expecting a baby, she could not follow us into the high mountains and had been obliged to spend a rather boring summer in Denver. The experiment that we had in mind was only meant to check whether the mean free path of mesotrons, did, in fact, increase with increasing momentum, as predicted by eqn (3.2).

The experiment consisted of a comparison between the mean free path before decay of two groups of mesotrons of different average momentum. The first group, selected by an anti-coincidence circuit, was formed of mesotrons, which, after traversing a shield (S) of 118 g/cm^2 of lead, came to rest in a second shield Σ of 115 g/cm^2 of lead; (as explained in Chapter 2, by the use of the anti-coincidence method it was possible to measure the flux of these mesotrons with the necessary accuracy). The second group consisted of mesotrons capable of traversing both lead shields (see Fig. 3.3). Clearly, the average effective momentum is greater for the mesotrons of the second, more penetrating, group than for those of the first group.

That summer we made our measurements at Denver and at Echo Lake. We compared the mesotron intensity observed at Denver with that observed at Echo Lake under an iron shield of a thickness equivalent to that of the air layer between the two stations as far as ordinary energy losses were concerned. Thus, if mesotrons had been stable particles, the two measurements should have given identical results. Instead, the intensity observed at Denver without the iron shield was considerably smaller than that observed at Echo Lake under the shield, a further proof of the anomalous absorption of mesotrons in air, and therefore of their decay in flight. Moreover the anomalous absorption was much greater for the less penetrating mesotrons. More precisely, we found that the mean free paths before decay of the two groups of mesotrons were, respectively, 4.5 kilometers and 13.3 kilometers. This meant that, indeed, the mean free path of mesotrons before decay increases with their momentum, as predicted by the relativity theory.

In 1941 Nora and I were back in Colorado for the third time. That summer six-month old Florence was with us. Nora and the child stayed at a small inn not far from Echo Lake and, once in a while, came to visit us where we were working. Our team included

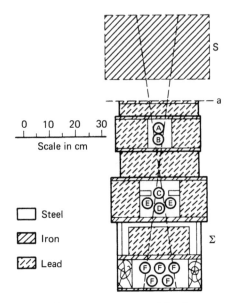

Fig. 3.3. Experimental arrangement used in the summer of 1940 to demonstrate that the mean free path before decay of mesotrons increases with increasing momentum. Two groups of mesotrons were selected: a 'hard' group formed by mesotrons which discharge counters A, B, C, D, F (identified by the coincidences ABCDF) and a 'soft' group of mesotrons which, after traversing counters A, B, C, D, stop in the absorber Σ and thus fail to discharge counters F (identified by the *anti-coincidences* ABCD–F). Counters E in anti-coincidence were used as a protection against spurious events. From an article by B. Rossi & D. Hall in *Physical Review*, 59, 233 (1941).

Kenneth Greisen, then a student, whose name will appear again in these notes. It also included three scientists from the local universities: Joyce Stearns and Darol Froman from the University of Denver, and Phillip Grant Koontz of the University of Colorado in Boulder.

That summer, with our experiment we wanted to go beyond the qualitative result obtained the year before by attempting to quantitatively verify the proportionality relation between momentum and mean free path before decay expressed by eqn (3.2). For this purpose it was necessary to measure with a fair degree of accuracy the average effective momenta of two different groups of mesotrons. This measure was possible only if each of the two groups was formed by mesotrons with momenta within sufficiently narrow limits. Because of the one-to-one relation between momentum and penetrating power, we could select the two groups of mesotrons with limited momenta by singling out mesotrons with a correspondingly limited penetrating power.

Accordingly, we recorded, by the anti-coincidence method, the mesotrons which, after traversing a given absorber (different for the two groups) came to rest in a second absorber. The computed effective momenta of the two groups turned out to be in the ratio of about 1.89. We then measured the mean free paths of the mesotrons before decay by observing their anomalous absorption in the air layer between Denver and Echo Lake. The values of the mean free paths of the two groups turned out to be in the ratio of 1.99. Therefore, within the experimental and computational uncertainties, the mean free path before decay of mesotrons was indeed proportional to their momentum, as predicted. Having reached these results, I felt that there was not much point in continuing mountain experiments. Thus, for the time being at least, the expedition of 1941 was my last visit to the Rocky Mountains.

Before turning to a different subject, I note that eqn (3.2) makes it possible to compute the mean life of mesotron at rest (τ_0), in terms of, the mesotron mass (m), and of p and l; these being, in principle, 'observable' quantities. However, because of the limited precision that can be achieved in their evaluation we can not expect to obtain more than a crude estimate for τ_0, which was found to be on the order of a few microseconds.

Here I must step back and inform the reader that our wandering had not come to an end in Chicago, and that, for one year now, we had been in a new place. To leave Chicago had not been easy. We were tired of moving around, and also we had grown fond of that windy city – sometimes torrid, sometimes frozen – a city which had given us shelter when we most needed it, where we had begun to learn about the country which was to become our new home.

Chicago had given us a great deal. At the University we had found a refined and active cultural center. The Science Division included many prominent scientists—scholars such as the physicist Samuel Allison and the chemist Robert Mulliken to name just a few. Like

Compton, all of them had done their best to make us feel at ease by opening the doors of their homes to us, by inviting us to join in their activities, by helping us in our practical problems, such as finding an apartment and buying a car. We also had friends outside the department: among them the writer Giuseppe Borgese (who had just married the young daughter of Thomas Mann), the Hungarian painter Gyorgy Kepes and his British wife Juliet, whom we were to meet again at MIT and who are still among our dearest friends.

But in Chicago I did not have a permanent position, nor did I earn a salary sufficient for our needs. Compton had tried to obtain a position of assistant professor for me, but had failed to do so because of the difficult financial situation of the University. Thus I had started a tour of the Midwest, visiting a number of universities, giving lectures, inquiring about the availability of an academic position. Everywhere I was received with great courtesy and what I felt was a sincere expression of regret that, because of financial straits, the University was not in a position to create a new professorship for me.

I am indebted to Hans Bethe for the solution of my problem. The death of a professor had created a vacancy in the Physics department of Cornell University. Hans, a member of that department, had suggested that I be invited to fill this vacancy. The suggestion had been received favorably, but the department wanted to meet me personally before making a decision; therefore I was invited for a visit. This was happening in the spring of 1940. I was about to go to Washington for the annual meeting of the American Physical Society and had decided to make a detour through Ithaca, the seat of Cornell University.

Nora and I left Chicago early in April, with a second-hand car which we had just bought. Our trip was a memorable adventure. At that time in Chicago one could obtain a driver's license without passing an exam; it was only necessary to state that one knew how to drive. In Italy, before departing, I had taken a few driving lessons; therefore I felt entitled to make the required declaration. But, in fact, I had only a vague notion of how to handle a car.

Our friends in Chicago were worried about our trip, and had recommended that we never exceed the speed of 45 miles per hour; thus Nora had kept her eyes riveted to the speedometer, shouting a warning whenever I exceeded, no matter how slightly, the prescribed limit. But 45 miles per hour was the maximum speed, not the average speed; because I was scared to pass other vehicles, and so I often remained for miles and miles behind trucks driving at 20 or 30 miles per hour. We crossed Cleveland in the rush hour. Then, turning north, we found ourselves in the midst of a snow storm. But eventually we reached Ithaca without any mishap.

At Cornell I met many members of the Physics Department and I gave a seminar. After the end of our visit, Nora and I continued our trip. At our arrival in Washington I was completely exhausted, but I had learned how to drive. Back in Chicago, while I was making preparations for the experiment in Colorado, I received the offer of a position of associate professor from Cornell University. Thus in the fall of 1940, upon our return from Colorado, we packed our belongings and left for Ithaca.

The town of Ithaca rises on a hill, overlooking Cayuga Lake, one of the five Finger Lakes which, long and narrow like the fingers of a hand, cut across the countryside of northern New York state. The view of the lake and a luxurious vegetation – gift of a damp climate – are attractive aspects of the place; despite frequent rains, summers in Ithaca are rather pleasant; not so the winters, cold and snowy. The University had rented a small cottage for us, some distance from the University and the business district, in an open and isolated area. In December, Florence was born, and I still remember the race, at night, to the hospital over icy roads, so slippery that, at all times, I was afraid of losing control of the car.

The first student who asked to work with me was Kenneth Greisen, whom I have already had occasion to mention. He became my first American pupil, and I could not have hoped for a more brilliant and conscientious collaborator. As it was already easy to predict, in a few years he would acquire a prominent position among American physicists.

One of the first fruits of our collaboration was a long article published in the *Review of Modern Physics* under the title: 'Cosmic-Ray Theory' (which came to be known as 'The Bible' in the cosmic-ray community!).

In the first part of this article we summarized the results of the theory of the electromagnetic interactions of charged particles and photons. In the second part we made an attempt to co-ordinate and to supplement the various results of the shower theory scattered in the scientific literature, and to exhibit them with equations and diagrams, in a form that was directly applicable to the interpretation of the experimental results. We then used our results to examine an important problem which, for some time, had received much attention, but had not yet found a solution. This problem concerned the origin of the 'soft' component of the local cosmic radiation. It was known that this component consists of electrons, with a small admixture of slow mesotrons. But it was not clear if all electrons were the products of secondary processes of mesotrons or if some had a different origin.

In an attempt to solve this problem, we computed, at various elevations, the ratio between the intensities of the 'soft' and 'hard' components of the cosmic radiation. The 'hard' component was known to consist of fast mesotrons. In the 'soft' component we included 'slow' mesotrons and electrons produced by the decay of mesotrons and by collisions of 'fast' mesotrons with air molecules. We took into account processes of cascade multiplication.

We then compared the results of these computations with measurements of the 'soft' and 'hard' components taken by Greisen in Ithaca (259 meters above sea level) and at various heights in Colorado.

From this comparison it turned out that the intensity of the 'soft' component was greater than was to be expected if all electrons had been produced by mesotrons. Moreover it was found that the excess of the observed over the computed intensity of the 'soft' component increased rapidly with increasing height. Thus, while in Ithaca the experimental and theoretical results were not very different from one another; at the top of Mt Evans the intensity of the 'soft' component was almost twice that predicted. Therefore, it was clear that the local cosmic radiation included not only electrons produced by mesotrons, but also electrons of a different origin. For want of a better explanation, we suggested that they might be the result of cascade multiplication in the atmosphere of electrons or photons present in the primary cosmic radiation. Five years went by before the discovery of neutral π-mesons provided the correct explanation of our observations.

I still felt that my commitment to the elucidation of the phenomenon of the radioactive decay of mesotrons had not been entirely fulfilled. The mountain experiments had removed any doubt about the existence of this phenomenon; however, as I have already noted, they only provided an order of magnitude for the mean life before decay of mesotrons at rest. Apart from the limited accuracy of the measurements, the reason for the uncertainty was that, as shown by eqn (3.2), the experimental quantity l on which the evaluation of τ_0 rested, is not directly related to τ_0, but is related to the ratio τ_0/m, the mesotron lifetime (τ_0) to its mass (m), a quantity, the latter of which at that time was only known approximately.

These considerations prompted me to plan an experiment entirely different from the previous mountain experiments, and designed so as to provide an accurate measure of the mean life of mesotrons at rest, independently of their mass.

The experiment was based on the observation of the decay electrons of mesotrons brought to rest in an absorber. A few attempts at measuring the mean life of mesotrons by a similar method had been made previously. The only experiment which had produced a significant result had been performed by Rasetti. Using three coincidence circuits with different resolving powers, he had recorded the coincidences between pulses signalling the arrival of mesotrons upon an absorber and pulses signalling the delayed emission from the absorber of their decay electrons. Under the assumption of an exponential decay curve, he had obtained a value of 1.5 microseconds for the mean life of a mesotron with an uncertainty of 0.3 microseconds. With the experiment I had in mind I was planning to achieve a much higher degree of accuracy. I also intended to actually measure the decay curve of mesotrons (i.e., the time dependence of their rate of decay).

For my experiment, I asked the co-operation of Norris Nereson, then a graduate student. He was already an experienced experimentalist, and his contribution proved to be a major factor in the success of our venture. The 'heart' of the experiment was a very precise 'electronic chronometer' which, to my knowledge, was the first of the time-measuring devices, later known as *time-to-pulse height* converters. The experiment consisted of measuring, by means of this chronometer the time interval between the arrival of a mesotron upon an absorber, where it came to rest, and the emission from the absorber of the electron arising from its decay (see Fig. 3.4 for further details).

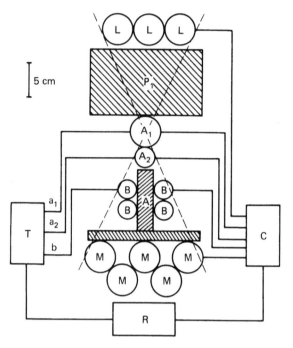

Fig. 3.4. Experimental arrangement used for the measurement of the decay curve of mesotrons at rest. An electronic circuit (C) records the anti-coincidences produced by mesotrons which, after traversing counters L, A_1, A_2 and the screen P_1, come to rest in the absorber A and thus fail to discharge counters M. Counters B are discharged by the electrons produced by the decay of these mesotrons. The electronic 'clock' (T) measures the time intervals between the arrival of mesotrons on the absorber (A) and the emission of their decay electrons. From an article by B. Rossi & N. Nereson in *Physical Review*, **62**, 417 (1942).

Briefly our 'chronometer' operated in the following manner: an electric pulse signalling the arrival of a mesotron, started an electric current, which gradually began charging a condenser. A second pulse, signalling the emission of the decay electron, stopped the current and discharged the condenser. The voltage of the condenser at the moment of the discharge was a function of the time interval between the two pulses. We used a quartz oscillator to measure this function exactly and were confident that the overall accuracy thus obtained amounted to better than 0.2 microseconds. (As the nation was at war and we thought that perhaps the 'chronometer' might have applications in some war-related research requiring the measurement of very short time intervals, we did not publish the description of our electric clock until after the end of the war—a minor example of the self-imposed secrecy rule that was common practice among scientists in those years.)

Using absorbers of different materials, we measured about 3000 decay processes with our chronometer, with delays greater than 0.8 microseconds (this being a lower limit set by the spontaneous delays of the G.M. counters). The decay curves obtained from these measurements had exactly the same exponential shape that characterizes the decay curves of ordinary radioactive substances (see Fig. 3.5). This result was not unexpected. Just the same, it was gratifying, for it provided a definitive proof of the validity of our experiment, and, I would like to add, it had some aesthetic value.

From the decay curve, the mean life of mesotrons at rest was found to be 2.15 microseconds, with an uncertainty of about three percent. Years later I was pleased to learn, from a letter of Luis Alvarez, that within its small experimental uncertainty, our value of the mean life agreed with the very precise value of this quantity (1.88=0.001 microseconds), obtained with measurements on mesotrons produced by man-made accelerators.

Here I wish to recall that, after the war, I learned that while Nereson and I were working in our comfortable laboratory at Cornell University, two of my Italian colleagues, Marcello Conversi and Oreste Piccioni, working in Rome in a cellar to escape the German occupation army, had succeeded in performing an experiment aimed, like ours, at a measurement of the mean life of mesotrons at rest. Using a method similar to that used by Rasetti, they found a value of 2.3 microseconds for the mean life, with an uncertainty of about ten percent. Within this uncertainty their result was in agreement with ours.

Our experiment was completed in the summer of 1943. We were now in the midst of the war. For some time, I had been working at problems of instrumentation for the development of radar, on a contract from the Radiation Laboratory of MIT. This work required frequent trips to Cambridge, Massachusetts. At that time the connections between Ithaca and Cambridge were not easy: by

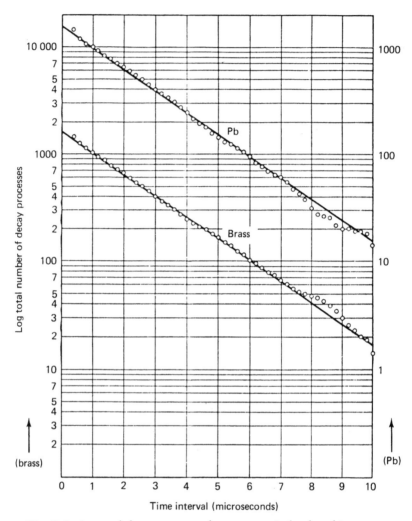

Fig. 3.5. Integral decay curves of mesotrons in lead and in brass. The abscissa is the time interval between the arrival of a mesotron on the absorber (A) and the emission of its decay electron. The ordinate is the logarithm of total number of decay processes whose delay is greater than the one shown by the corresponding abscissa. Note that a straight line on a semi-logarithmic plot corresponds to an exponential curve on a linear plot. From an article by N. Nereson & B. Rossi, in *Physical Review*, **64**, 199 (1943).

bus to Syracuse, by train from Syracuse to Albany, and a long wait at Albany for a train that would take me to Boston and Cambridge.

For me, there were redeeming aspects: the chance of learning about a new, highly sophisticated technology of unique value for the war effort, a technology which was responsible for thwarting the Nazi invasion of England; the gratification of being in a position of making a small contribution to its development (I still conserve a patent related to one of these contributions); and the occasion for meeting interesting people both inside and outside of the laboratory. At that time Gaetano Salvemini lived in Cambridge supported by a modest grant from Harvard University. We often had lunch together at a cafeteria in Harvard Square. His conversation, showing a clear, deep knowledge of the political situation on one side, and a heart-felt compassion for human plights on the other, was such as to make me forget the quality of the food that was being fed to us. Worse off was Nora, who was left alone, with a child of less than two years, in our isolated cottage, often half buried in the snow, with an ancient heating system and with a gasoline supply hardly sufficient for driving to town once a week.

In the meantime came the call to Los Alamos, a call that recast my life, as it did the lives of so many other people. There followed days of doubt, of anxiety which added to the worries for our families left in Italy. I remember finding some relief in the thought that the research project which I had started four years earlier had helped to clarify some aspects of the physical reality; a modest accomplishment, to be sure, but one that would survive all human upheavals. (This is the strength or, if you wish, the weakness of the scientist.) Our experiments had produced the final proof of the radioactive instability of mesotrons, had established a precise value for its mean life, and as a by-product, had verified the time dilation of moving clocks predicted by the relativity theory.

Today, when I think of the kind of research activities in which I and other scientists were engaged in those years, I am overtaken by a feeling of unreality. How is it possible that one could obtain results affecting problems of fundamental significance by means of experiments of an almost childish simplicity, experiments costing just a few thousand dollars, requiring only the help of one or two students?

In the decades that have elapsed since then, the study of elementary particles has moved from cosmic rays to the great accelerators. These machines have provided the scientists with research facilities of a power and sophistication previously undreamt of. All of us, of the old generation, have witnessed with great admiration this extraordinary technological development. And yet, deep in our hearts, we feel a persistent nostalgia for that era which for lack of a better expression, I shall call the 'Age of Innocence of the Physics of Elementary Particles'.

4

Los Alamos

(1943–46)

In the early morning of July 19, 1943, I was deposited by the *Superchief* – the transcontinental train, pride of the Santa Fe Railroad System – at the station of Lamy, a small village in the middle of nowhere. A car was waiting for me. We left immediately for Santa Fe, the capital of New Mexico, some fifteen miles to the north. In Santa Fe the driver took me to one of the small offices which opened their doors under the arcades of a narrow street, in the center of town. Alone in the office sat a middle-aged lady, who kindly rose to greet me and to welcome me to New Mexico. She was Dorothy McKibben, who had the responsibility of keeping a record of all persons *en route* to Los Alamos. She knew that up there very important and highly secret work was under way. But she did not know then, neither did she want to learn until the end of the project, what the precise nature of this work was.

Upon leaving Santa Fe we drove north on the federal highway headed toward Taos. The road ran over a sandy plateau sparsely dotted with small bushes of junipers and dwarf pine trees. At the horizon, row after row of low mountains, dark and sharply outlined against the sky in the foreground turned progressively paler, becoming almost evanescent with increasing distance. Shortly after the village of Pojoaque we turned west on a narrow dirt road which descended gradually into the Rio Grande valley, and after crossing the river, started climbing rapidly, with several switchbacks, until it reached Los Alamos, on top of a mesa, 2400 meters above sea level in the Jemez range.

The recollection of that trip is still vivid in my mind. I had never been to New Mexico before, and I felt as if I were in a fairyland. We passed some rock formations of fantastic shapes. From the bridge on the Rio Grande I saw a large stretch of Indian land with herds of horses grazing by the river; in the background, a mesa of a dark color and of an unusually regular trapezoidal shape (as I learned from the driver, it was the sacred *Black Mesa*, used by the Indians of the nearby Pueblos for their religious ceremonies). And high above, a luminous light blue sky, alive with passing small white clouds.

The invitation to join the Los Alamos project had been brought to me by Hans Bethe early in July. The days that followed this invitation were among the hardest of my life; without being told, I could easily imagine what the Los Alamos project was about, and I was loath to have any part in the development of such a deadly device as the fission bomb was expected to be. On the other hand, I like many others, was terribly worried by the likelihood that in Germany, where fission had been discovered, work on the bomb might be advancing at a fast pace. Finally, having resigned myself to the fact that neither by accepting nor rejecting the Los Alamos request could I escape a heavy responsibility, I decided that my choice could not be based on anything else but the need to fight the immediate danger.

I clearly remember my feelings when I decided to go to Los Alamos. I was hoping that our work would prove that the fission bomb was not feasible. However, I had also reached the conclusion that if, on the contrary, the bomb was feasible we must make sure, at all costs, that Hitler did not have it before we did. Having reached a decision, I felt an urgency to act upon it, and so I left for Los Alamos alone, as soon as I could. A short time later, I would take off a few days to go back east and return to Los Alamos with Nora and Florence.

Because of its secluded location on the one hand, and the vicinity of a city on the other, Los Alamos (*the poplars*) had been an ideal choice for a project requiring the strictest secrecy as well as easy access to the outside world. Before being requisitioned for the project, Los Alamos had been the home of a boarding school for children of wealthy families. Inherited from the school was a large, handsome log building (the *Lodge*), now used as a restaurant, a lodging for visitors, and a meeting place. There were also a few small stone houses (which became known as the *bath-tub-row*, being the only dwellings on the hill equipped with bathtubs rather than showers). These were assigned as residences for the most *important* members of our community. For the rest of us, the army engineers had hurriedly built some wooden structures, more barracks, actually, than houses, each with two or four apartments.

Much has been said and written about the discomfort of life in Los Alamos. In my opinion these complaints are not warranted. True, our living quarters were somewhat primitive and streets were often full of mud or dust; true, water (pumped up from the Rio Grande) was scarce; there was only one grocery store, often not too well supplied; true, we could not move more than forty miles from the site without special permission. But it was wartime, and how insignificant these minor inconveniences were compared with the tragedies in Europe. Besides, there were redeeming features. The invigorating mountain air, a dry climate, quite variable but rarely too hot or too cold, and the occasional long walks in the mountains, among woods of pine trees and aspens which turned bright gold in the autumn. There was the opportunity of meeting interesting people belonging to the permanent staff or coming up as visitors. And there was the discovery of the Indian culture, so different from ours, yet so appealing, so strongly conditioned by aesthetic values. A few miles from Los Alamos, by the Rio Grande, lies San Ildefonso, perhaps the best sample of Indian traditional town planning in that region. San Ildefonso was also the home of Maria Martinez, whose black pottery, fired in the open in fascines of twigs and horse manure, can today be admired in museums and private collections. For us, finally, the communal life at Los Alamos was a welcome change from our isolated existence in Ithaca.

While I am on the subject of life in Los Alamos, I would like to relate a small episode of which Nora was the heroine and which became part of the local folklore. At our request, we had been transferred from an apartment in the center of the compound to one facing a small canyon which separated the living section from the technical area. It was an ideal location because of the view from our windows and because the canyon was a wonderful playground for the children, as well as a pleasant picnic site. One morning we were awakened by the din of bulldozers at work in front of our house. We looked out; the canyon had disappeared and so had the little hill behind it, which simply had been pushed down into the canyon. (We were told later that this had been done to enlarge the technical area.) As if this was not enough, a bulldozer was systematically pulling down all the trees in front of our house. Only one tree was still standing, a very large and beautiful pine tree. Nora rushed out of the house and placed herself in front of this tree, determined to save it. The bulldozer approached and, shouting over the noise of the engine, the driver ordered Nora to clear out, but she refused to move. After shutting off the engine, the driver explained that he had been instructed to tear down all trees near the houses, as a precaution against a fire. Clearly 'our' tree did not represent any fire hazard, so Nora answered that she would move only when she saw a written order specifically concerning the tree in question. Eventually the bulldozer withdrew and was not seen again, and the tree was saved.

The Los Alamos laboratory had opened officially in April 1943; its director, as is well known, was Robert J. Oppenheimer. I had known Oppenheimer only slightly before going to Los Alamos, but, once there, it did not take long for me to become impressed by his truly exceptional qualities as a scientist and leader. At all times he would keep himself fully informed of everything that was going on in the laboratory and would know it sufficiently well to offer constructive criticism and advice. While keeping rigid control over the general direction of the project, he somehow succeeded in doing so without interfering with the personal initiatives of the scientists.

My admiration for Oppenheimer did not close my eyes to the sometimes disconcerting complexity of his character. Except perhaps for some of his most intimate friends, I felt that none of us really knew him entirely, a feeling due in part to some occasional episodes of puzzling behavior. But whatever the peculiarities of his character were, Oppenheimer was eminently successful in keeping together, for the duration of the project, a group of scientists very independent by nature and training, and to ensure that they all worked jointly toward a pre-established goal. This was a difficult task, requiring not only scientific leadership, but also the human sensitivity needed to ward off possible causes of disruption and to recognize emotional problems that were bound to arise in a community such as that in Los Alamos.

Los Alamos was part of the Manhattan Project, the organization which co-ordinated all activities related to the production of a nuclear bomb. Like the rest of the Manhattan Project, Los Alamos was subject to military control. The person in charge was General Leslie Groves, about whose merits and shortcomings scientists are sharply divided. However, I do credit him with a most important act of wisdom: his decision not to interfere with the activities of the scientists formally under his command, but to let them proceed in their work as they saw best.

At first there was an attempt to enlist all of the Los Alamos personnel in the army, the alleged purpose being that the military discipline would provide maximum protection for the secret of the bomb. This action would have seriously hampered the research work by obstructing the free flow of information between scientists at all levels. Therefore the proposed measure was resolutely opposed by the Los Alamos staff and was soon abandoned.

Even without a military regime, rigorous security measures were in force in Los Alamos. The nature of the project was jealously kept secret not only from all outsiders (one of the rumors going around in Santa Fe was that Los Alamos was a home for pregnant WAAC's – Women's Army Auxiliary Corps), but also from all workers in Los Alamos who did not need this information in order to perform their duties. Even wives were supposed to be kept in the dark but it proved impossible to prevent at least some of them from guessing the general purpose of the work. The word 'bomb' could not be uttered in Los Alamos, and the word 'gadget' was used in its place.

Speculations about the possibility of building a nuclear bomb had started soon after the discovery of fission in 1939. As is well known, fission is the name of a process whereby the nucleus of a heavy isotope (e.g., one of the uranium isotopes), when struck by a neutron, splits into two lighter nuclei and ejects a few neutrons. It so happens that the combined mass of the fission products is less than the mass of the original nucleus. The mass which is thus lost, call it Δm, is converted into energy (E), according to Einstein's law:

$$E = \Delta m \ c^2$$

where c is the speed of light. This is the energy which in a nuclear power reactor is released gradually, under controlled conditions. This is the energy which, when released explosively, has the deadly effects of a powerful bomb. In either case, the mechanism responsible for the energy release is the *chain reaction*, a process in which neutrons produced by fissions in a mass of fissionable material proceed to produce further fissions in the same mass.

If the number of fissions in a given generation is less than the number of fissions in the preceding generation (a situation referred to as *subcritical*) the nuclear activity of the fissionable mass will decrease with time. If, instead, the number of fissions increases from one generation to the next (known as *supercritical*), the nuclear activity will increase with time (the chain reaction may then be called *divergent*). If the rate of increase is sufficiently fast, very large amounts of nuclear energy will be released in a short time interval and an explosion will result. Which of the situations described above will actually prevail depends on the number of neutrons that are 'lost' either because they escape from the fissionable mass, or because they undergo nuclear absorption without producing fissions.

In a report distributed to the laboratory staff, and known as the *Los Alamos Primer*, Robert Serber, a member of the Theoretical Division, described in some detail the problem of the nuclear bomb, as it was perceived at the start of the Los Alamos project. When the *Primer* was written, the properties of an isotope capable of producing an explosive chain reaction had been clearly enunciated. They included the number and energy spectrum of the neutrons produced in the fissions, the fission cross-section of the isotope, and the probability for the fission neutrons not to be captured without producing further fissions. At that time, only one isotope was definitely known to possess these properties. This was the rare uranium isotope of mass number* 235, (U235) which is present in the proportion of 0.7 percent in natural uranium (the rest is almost entirely U238).

It had been predicted, however, that the plutonium isotope of mass number 239 (Pu239), which so far had been produced only in microscopic quantities, should have properties similar to those of U235. The most serious doubt about the possibility of using Pu239 as a bomb material concerned the number of neutrons generated in the fission of this isotope.

* The *mass number* is the total number of protons and neutrons in a nucleus. It is *approximately* equal to the actual mass of the nucleus, in terms of the proton mass, the difference being due to the relativistic mass equivalent of the binding energy of protons and neutrons in the nucleus.

Anticipating a result to be obtained a few months later, I shall mention here that Emilio Segrè, from a direct comparison between the behavior of Pu239 and U235, found that the number of neutrons produced by the fission of Pu239 was actually slightly greater than the number of neutrons produced by the fission of U235. Thus, in principle, both U235 and Pu239 could be used to make a bomb. This was a result of great importance because Pu239 could be produced abundantly in the nuclear reactors, as soon as these became operational (the process being the absorption of a neutron and the emission of an electron from U238), whereas the extraction of the rare isotope U235 from natural uranium was a difficult and slow process hardly capable of producing a sufficient amount of bomb material in the timeframe required.

In planning the production of a nuclear bomb, the first problem to be faced was that of the purification of the bomb material, because it was essential to avoid the neutrons being absorbed by nuclei of foreign substances before they had a chance to produce further fissions. This was a problem for the radiochemists. Another condition that needed to be satisfied concerned the minimum number of neutrons which must be kept within the fissionable mass to allow a chain reaction to develop. Since neutrons escape more easily from a small mass than from a large one, it was clear that to reach a supercritical state it was necessary to use a fissionable mass greater than a certain amount.

A reason for concern was the large amount of heat developed in the early stages of a chain reaction. The consequent increase in temperature of the fissionable mass causes the mass to expand, and since neutrons escape in increasing numbers as the density of the mass decreases, eventually the thermal expansion will stop the chain reaction. The *Primer*, reporting the results of some previous computations, stated that this would happen about five-hundredths of a microsecond (5×10^{-8} s) after the beginning of the chain reaction. It was thus necessary that the reaction should develop sufficiently fast to ensure that a large amount of energy was released in this short time interval.

This was the reason why a process similar to that used in the Fermi pile (which had come into operation in December 1942) could not be applied for the production of a nuclear bomb. The nuclear fuel used in the Fermi pile had been natural uranium, but the active component had been the rare isotope U235. In order to produce a chain reaction, it was necessary that the neutrons arising from the fissions of the nuclei of U235 should go on to produce further fissions of the nuclei of the same isotope without first being absorbed by the much more abundant nuclei of U238. In the Fermi pile, this end was achieved by rapidly slowing down the neutrons so as to take advantage of the very large increase of the fission cross-section of U235 which takes place when the neutron energy is decreased. The slowing down, which was achieved by anelastic collisions of neutrons with carbon nuclei in a graphite moderator, required about one microsecond (10^{-6} s). Therefore the rate of growth of the chain reactions in the Fermi pile, and in all other reactors operating on a similar principle, was much too slow for the purpose of producing a nuclear explosion. To achieve this end it was necessary to use chain reactions in which the fission neutrons directly produce further fissions without the need of a previous slowing-down. In these reactions to be described as *immediate chain reactions*, the time interval between two subsequent generations is on the order of one-hundredth of a microsecond (10^{-8} s). Thus, these chain reactions will develop sufficiently fast to avoid premature termination by the thermal expansion. The available data on the nuclear properties of U235 and Pu239 indicated that both isotopes were capable of generating immediate chain reactions (this in fact, was the reason why they had been singled out as suitable isotopes for the production of a nuclear bomb). It was, of course, desirable to obtain experimental information on the rate of development of the chain reactions in these two isotopes (and I shall return to this subject later); however on the basis of existing knowledge one could be confident that thermal expansion would not be a serious obstacle to the development of fast chain reactions in U235 or Pu239.

More serious was the problem created by the unavoidable presence of stray neutrons: neutrons in the cosmic radiation; neutrons produced by α-particles arising from traces of radioactive substances; and neutrons produced by spontaneous fission. In order to ignite a nuclear bomb, it was necessary to bring a mass of fissionable material from a subcritical to a highly supercritical condition. During this transition, the fissionable material had, by necessity, to go through a weakly supercritical condition. If a stray neutron should initiate the chain reaction during this phase of the transition, the fissionable material would heat and expand. The chain reaction would thus be stopped when only a small fraction of the available nuclear energy was released, in which case no explosion would occur. In order to avoid such *predetonation*, i.e., such a premature initiation of the chain reaction, it was necessary to perform the transition in a very short time; the shorter the time, the more abundant are the stray neutrons. In the *Primer* it was stated that the information then available had placed an upper limit of 100 microseconds (10^{-4} s) on the acceptable duration of the transition. (As we shall see, this estimate was optimistic.) The development of a method for the fast assembly of a supercritical mass was a difficult problem, whose solution was a condition *sine qua non* for the success of the efforts for the production of a nuclear bomb. Los Alamos had been assigned the task of finding a solution to this problem.

Prior to the writing of the *Primer*, only one method had been suggested for the rapid assembly of a supercritical mass. This method (called the *gun method*) required the use of a gun of special design to shoot a bullet of fissionable materal against a target also made of fissionable material. Separately, the bullet and the target would each have a subcritical mass. Together they would form a supercritical system. Soon afterward, however, Seth Neddermeyer, a physicist from the California Institute of Technology, came up with a different suggestion based on the consideration that neutrons remain more effectively trapped in a mass of fissionable material the more compact the mass. Therefore, he argued, it must be possible to rapidly bring a mass of fissionable material from subcritical to supercritical condition by increasing its compactness with the help of a high explosive.

The general principle underlying this method (later called the *implosion method*) was soon concretized into the specific proposal of a bomb consisting of a spherical shell of fissionable material surrounded by an explosive. When this was detonated, the shock wave would compress the shell changing it into a solid sphere, thereby increasing the compactness of the fissionable material. The question of whether to rely on the gun method or the implosion method for igniting the nuclear bomb was debated for a while. On the one hand, theoretical considerations suggested that the implosion method should be able to produce a supercritical assembly more rapidly than the gun method. On the other hand, the development of the implosion method involved many more serious problems than the development of the gun method—problems whose solutions were likely to require more time than was available. Remember that implosion was a 'new' phenomenon, one which had never been studied either experimentally or theoretically. Moreover, preliminary experiments by Neddermeyer himself had given results that were both encouraging and worrisome. Examination of imploded hollow cylinders had shown that implosion had actually produced the predicted increase of compactness, but it had also brought to light certain faults (asymmetries, fractures, jets) which needed to be corrected before a supercritical condition could be achieved.

Because of these difficulties, and because from the available information about the abundance of stray neutrons, it appeared that, both in the case of U235 and Pu239, the speed of assembly provided by the gun method was sufficient to avoid predetonation, it was decided to give the gun method top priority and to place implosion on the back burner.

When I first arrived at Los Alamos, Oppenheimer asked me to form a group that would help design and build the instruments needed by the various experiments being planned in the laboratory. This was an easy assignment for me because, thanks to my previous work in cosmic rays, I had acquired a certain familiarity with radiation detectors and associated electronic circuits. It so happened that at Los Alamos there was already a group with an assignment similar to ours. This group was headed by Hans Staub, a Swiss physicist. After comparing our programs, Hans and I decided to join forces by merging our two groups into one, which became known as the *Detector Group*, or group P-6 (group No. 6 of the Physics Division, headed by Robert Bacher of Cornell University). This was a fortunate decision because despite the difference between our temperaments – or perhaps because of this difference – my association with Staub turned out to be one of the most gratifying aspects of my life and my work in Los Alamos. Our group, quite small at first, grew rapidly until it included some twenty people, many of them in their twenties, just out of college. I gratefully recall that to a large extent it was the intelligent and tireless work of these young people which enabled us to perform the various tasks that we were assigned in record time.

During the first months, in compliance with our assignment, we developed a number of electronic circuits and counting devices, sometimes in collaboration with people in the electronic group, and in particular with Mathew Sands, still a student, but already an electronic wizard. But the most important result of our work on instrumentation for the general use of the laboratory was the development of the *fast ionization chamber*. The motivation of this project was the expectation that the experimental program of the laboratory would require radiation detectors with the following properties: (1) a large sensitive area (in order to detect small radiation fluxes); (2) a response proportional to the energy released in the detector by the radiation, and (3) the capability of accurately following fast changes in the intensity of the radiation.

None of the detectors known at that time had all of the three properties mentioned above. True, the ionization chamber was a proportional detector and its sensitive area could be increased more or less at will; however it was generally believed to be an intrinsically slow detector, incapable of responding to rapid changes of an ionizing agent. Hans and I, feeling that this belief was not based on any sound experimental or theoretical evidence, took it upon ourselves to examine afresh what actually happens when ionizing radiation enters a chamber. On the basis of this study we prepared, for the use of the laboratory, a brief but accurate description of the operation of the ionization chamber, with a clear specification of the conditions in which the chamber would behave as a fast detector. I shall sumamrize our report here.

As is well known, the ionization chamber consists of a metal vessel filled with a gas and containing two electrodes held at different electric potentials. (One of the electrodes might be the metal wall of the container.) Radiation that crosses the chamber will ionize the molecules of the gas extracting an electron from each of them and leaving behind a positively-charged ion. In some cases the electrons are promptly captured by the neutral molecules changing them into negative ions. In other gases, the electrons remain free.

Free electrons have a much higher mobility than positive or

negative ions. In the electric field produced by the potential difference between the two electrodes they move rapidly toward the positive electrode (the *anode*). By an effect known as *electrostatic induction* the moving electrons produce an impulsive change of potential on the anode. This begins when the electrons are set free by the ionizing agent, and terminates when the electrons reach the anode. The actual shape of the pulse depends on the geometry of the chamber. In any case, because of the high mobility of free electrons, the electric pulse has a very short duration, which means that the chamber behaves as a fast detector.

It follows that, in order to obtain a fast response, it is necessary to insure that the gas does not capture the electrons but allows them to remain free. The operation of the chamber can thus be optimized by carefully choosing a gas where electrons would have the highest mobility. Starting from these results, Hans Staub and another of our co-workers, James Allen, examined the behavior of electrons in a number of gases and gas mixtures. Of all samples examined by them, the one in which electrons were found to have the highest mobility was a mixture of ninety percent argon and ten percent carbon dioxide. The pulses recorded with this mixture had a duration on the order of microseconds.

The fast ionization chamber became a basic tool for the experimental program of Los Alamos. In the words of David Hawkins, the official historian of Los Alamos: 'The development by the detector group and the electronic group respectively of new counting techniques involving electron collection and new fast amplifiers caused a minor revolution in the counting techniques and electronic equipment used by the Physics Division.'

While continuing its activity in support of the general experimental program of the laboratory, our group soon became engaged in some experiments of its own. For the most part, these were routine experiments, intended to provide useful information for the Los Alamos project, but they did not involve any new ideas or any new technology. Rather than dwelling on the routine activity of the group, I would like to briefly describe an experimental program which is of some interest because of its conception and its applications. This program aimed at developing a new method for the evaluation of the rate of increase of the activity of a divergent chain reaction.

Attempts had been made to theoretically estimate this quantity, but because of the great complexity of the computations and the uncertainty in the exact values of the physical quantities on which they were based, we could not rely entirely on the theoretical results in the absence of experimental verification. Of course, a direct measurement of the rate of growth of an explosive chain reaction had to wait until a bomb was actually exploded (and, as we shall see, I carried out such a measurement on that occasion). But, in the meantime, it was possible to test the validity of the theoretical predictions by some laboratory experiments. The underlying reasoning was that basically the same theory applies to both supercritical and subcritical systems. Therefore if the rate of decay of the nuclear activity of a slightly subcritical system was measured, and the measurements agreed with the theoretical predictions, one could be confident that the theory would also give the correct result for the rate of growth of the activity of a supercritical system. The most obvious method for measuring the decay of the activity of a subcritical system was to irradiate the system with short pulses of neutrons and then observe the decay of the activity induced by these pulses. Experiments by this method were actually performed by means of a pulsed cyclotron. But this was a rather cumbersome procedure, which required access to a cyclotron.

I learned about the problem shortly after my arrival in Los Alamos. It occurred to me that the desired information could be obtained by a much simpler, although conceptually more sophisticated method than that of the pulsed cyclotron. My method was based on the realization that in a slightly subcritical system (as in a supercritical system) fissions do not occur at random but appear in chains. It was possible to construct the decay curves of the individual chains by recording, with an appropriate detector, the neutrons emitted by the fissions in these chains. Since the development of a chain is a random process, the decay curves of individual chains would be different from one another. However, intuitively, I thought that by adding up the decay curves of a large number of individual chains, one would obtain the decay curve of the nuclear activity of the whole system. (Later a rigorous treatment of the problem by Richard Feynman showed that my intuition had been correct.)

Early in August 1943 I met with some of my friends in the

theoretical division (Hans Bethe, Edward Teller, and Richard Feynman) and discussed my plans with them. Their positive reaction encouraged me to try the experiment. I asked Nereson (who had followed me from Cornell) to prepare the experimental equipment. This was completed in August 1944. It involved the use of two neutron counters and operated in the following manner. The first counter, when struck by a neutron, produced an electric pulse which opened a succession of narrow 'gates' with progressively longer delays. A coincidence circuit then recorded the coincidences between these 'gates' and the pulses of the second neutron counter. After some time the total numbers of coincidences in the various 'gates' were plotted against their delays. The curve thus obtained was the average of the decay curves of the individual chains and was therefore the decay curve of the nuclear activity of the system.

The method which I have described was applied for the first time to a measurement of the decay curve of the activity of the so-called *water boiler*, a nuclear reactor built in Los Alamos, which used uranium enriched with the U235 isotope as a fissionable material, and water as a moderator to slow the neutrons. It was reassuring to find that the experimental results were in agreement with the theoretical predictions. Later the method, that became known as the *Rossi method*, was applied in several other experiments. But I did not participate actively in the continuation of this program because in the meantime, I had become involved in the problem of the implosion.

Despite the decision of using the gun method for firing the nuclear bomb, interest in the implosion method was still alive. In September 1943 it was brought into sharper focus by a visit of John von Neumann, the renowned Hungarian scientist, now at Princeton University, who, during the war, applied himself to problems concerning high explosives. During the following winter, the Los Alamos theorists had initiated a program to study the physics of implosion. To simplify the computations, they had assumed that the imploded object was in the shape of a spherical shell and that its implosion was produced by a spherical convergent explosive wave. Despite the assumption of spherical symmetry the study required very complex numerical computations which could not be com-

Plate 1. Accetri from a drawing by Meta Cohn-Hendel. In the background, the 'Giotello'.

Plate 2. Antonio Garbasso.

Plate 3. With Gilberto Bernardini, on the terrace of the Physics Institute.

Plate 4. On the steps of the Physics Institute. From left to right: Giuseppe Occhialini, Gilberto Bernardini; in the front, the author; in back, a technician.

Plate 5. At lunch in the Institute. From left to right: Lorenzo Emo-Capodilista, Beatrice Crino, Gilberto Bernardini, Attilo Colacevich, Daria Bocciarelli.

Plate 6. At work in the laboratory of the Institute. I had not yet bothered to learn how to work the power supply. Thus, I used a number of dry-cells connected in series to provide the voltage needed for the operation of the G.M. counters.

Plate 7. Daria building a G.M. counter.   Plate 8. Walter Bothe.

Plate 9. With Hans Geiger in Tubingen, summer of 1930.

Plate 10. The Conference on Nuclear Physics, Rome, October 1931. Among the participants: 1, Heitler; 2, Stearns; 3, Debye; 4, Richardson; 5, Millikan; 6, Compton; 7, Curie; 8, Marconi; 9, Bohr; 10, Bothe; 11, Rossi; 12, Sommerfeld; 13, Corbino; 14, Persico; 15, Perrin; 16, Fermi; 17, Ehrenfest; 18, Maiorana; 19, Garbasso; 20, Blackett; 21, Brillouin; 22, Heisenberg.

Plate 11. With Robert Millikan and Arthur Compton at the Rome Conference.

Plate 12. With Enrico Fermi.

Plate 13. After the Rome Conference, Lise Meitner and Walter Bothe stopped in Venice for a short visit. Here I am with them on the beach of the Lido.

Plate 14. With Erik Regener and Lise Meitner on the Bodensee (Lake of Contance) in the motorboat used by Regener for his underwater measurements of cosmic rays. Regener had named his boat *Undula* to signify his belief in the wave theory. I had tried to convince him that cosmic rays were of a corpuscular nature. Somewhat short of arguments Regener had answered smiling, 'But if you are correct, I would have to rename this boat *Corpusel*, and you will agree that it does not sound as nice as *Undula!*'

Plate 15. The new Physics Institute of the University of Padua.

Plate 16. The student laboratory of the Physics Institute.

Plate 17. The marketplace of Asmara.   Plate 18. A native woman of Asmara.

Plate 19. Negroid population of the interior.

Plate 20. The *Ascari* of the Army Engineers building a cabin for our experiment.

Plate 21. The cabin with the tent providing protection against the equatorial sun.

Plate 22. The cosmic-ray 'telescope' in the cabin.

Plate 23. Sergio De Benedetti.

Plate 24. Copenhagen. In front of the Bohr Institute; at the center Niels Bohr and George Hevesy.

Plate 25. The cosmic-ray symposium at the University of Chicago in 1939; at the center, Victor Hess. Among the scientists to the right: Vallarta, Compton, Bethe; on the left, Bothe, Heisenburg, Swann.

Plate 26. Nora and the author with the mother and the wife of Arthur Compton, at Otsego Lake.

Plate 27. The bus at
Echo Lake.

Plate 28. With Nora at
Echo Lake.

Plate 29. The cabin on one of the peaks of Mount Evans.

Plate 30. With Nora, on the second peak of Mount Evans.

Plate 31. At work inside the bus parked on the saddle at the top of Mount Evans.

Plate 32. Colorado, 1940. David and Jane Hall with Winston Bostich.

Plate 33. Colorado, 1941. From the left: Kenneth Greisen, a student, Betty Greisen, Phillipp Koontz.

Plate 34. The Camel Rock: a sandstone formation along the road from Sante Fé to Los Alamos. (Photo; Frank Rossi)

Plate 35. The view from the bridge on the Rio Grande. The land, partly farmed in corn and beans, partly kept fallow as a grazing ground, belongs to an Indian reservation. In the background, the sacred Black Mesa, the site of the Indian's religious ceremonies. (Photo; Frank Rossi)

Plate 36. Robert Oppenheimer at Los Alamos. (Courtesy of the Los Alamos National Laboratory)

Plate 37. The large *plaza* of the Indian pueblo of S. Ildefonso. At the far left the entrance of the *Kiva*, an underground chamber used by the Indians for religious ceremonies and as a meeting place for the elders of the pueblo. (Photo; Frank Rossi)

Plate 38. Maria Martinez; the Indian woman largely responsible for the revival and the refinement of the ancient ceramic art of her people. (Photo; Cradoc Bagshaw; printed in the book 'Maria' by R. L. Spivey, Northland Press, 1979)

Plate 39. One of the four trays of fast ionization chambers used for the RaLa experiment in Bayo Canyon. (Los Alamos files)

Plate 40. In Bayo Canyon. The radioactive RaLa source prepared in the laboratory of radiochemistry is taken by a truck to the firing site enclosed in a lead container. (Los Almos files)

Plate 41. At Trinity. The bomb was mounted on top of a 100-foot tall steel tower. The fast ionization chamber was suspended near the bomb. From the chamber the transmission line (the shorter of the two seen in the picture) slanted down to a point underground, then continued still underground to an underground shelter where the oscillograph and the other instruments were located. (Los Alamos files)

Plate 42. The explosion of the first nuclear bomb in the Trinity test. (Los Alamos files)

Plate 43. The log cabin used as a laboratory at Echo Lake.

Plate 44. At Echo Lake, in 1949. From left to right: Susan and William Walker of Cornell University, a Chinese visitor, John and Betty Tinlot, Bernard Gregory.

Plate 45. Nora, Frank and the fawn at Doolittle Ranch in 1949.

Plate 46. with Herbert Bridge in front of a large multiplate cloud chamber.

Plate 47. The Conference of the International Union of Pure and Applied Physics (IUPAP) at Bagnères de Bigorre in 1953.

Plate 48. Blackell and Leprince Ringuet at Bagnère de Bigorre.

Plate 49. Varenna, 1954. At the opening of the course in physics at the Villa Monastero. In front: the author, Bernardini, and Fermi.

Plate 50. The IUPAP conference in Moscow, 1955. In the center, D. Skobeltzyn.

Plate 51. Moscow, 1955. From left to right: Powell, Skobeltzyn, the author, Shapiro, Jánossy.

Plate 52. The string of balloons used by Robert Hulsizer in his search for electrons and photons in the primary cosmic radiation.

Plate 53. John Linsley at Volcano Ranch.

Plate 54. Volcano Ranch; the laboratory and the control room.

Plate 55. At Volcano Ranch. Nora with John Linsley, Livio Scarsi and their wives.

Plate 56. With Herb Bridge testing the plasma probe.

Plate 57. William Kraushaar and George Clark working at the detector for the X-ray experiment.

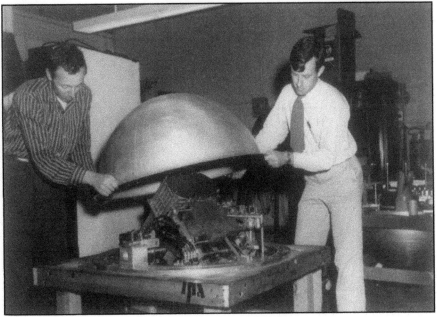

Plate 58. With Occhialini.

Plate 59. With Minoru Oda.

Plate 60. With Walter
Lewin in Rome.

Plate 61. Florence,
Bruno, Nora and Frank
in Los Alamos.

Plate 62. Linda with Pepper.

Plate 63. H. Yukava, Nora, Mrs Yukava, Bruno, and C. Pauli in Japan.

Plate 64. Bruno receiving one of his honorary degrees in Durham 1974.

Plate 65. Party for the 80th birthday of Bruno. Three generations of pupils: Clark, Bridge, Canizares, and Georgy and Juliet Kejes.

pleted until the arrival of the first IBM computers in the spring of 1944.

The results were of considerable interest. Among other things, they showed that implosion not only made the imploded object more compact, by changing a spherical shell into a solid sphere, but also caused a compression of its material. It was clear, however, that the theoretical results needed to be checked and supplemented by experiment. One concern was that the convergent spherical explosive wave postulated by theory was an ideal which one might possibly approach but never reach in practice. There were other worrisome problems that could be answered only by experiment. For example, the British scientist Sir Geoffrey Taylor, who had come to Los Alamos for an extended visit, had warned us that, even under ideal geometrical conditions of sphericity, the pressure exerted by a light material (the explosive charge) over a heavy material (the imploded metal) is apt to produce an instability which might have adverse effects on the implosion. This is the so-called *Taylor instability*, whereby if you turn over a glass of water, the water will spill although the air pressure should be sufficient to hold it in the glass – as, in fact, it does if, to prevent the instability, you place a sheet of paper over the glass.

Several experimental studies of implosion were planned and in part carried out while the theoretical work was still under way. Some of these used the intense light flash produced by the detonation of an explosive charge to photograph the interior of a hollow hemisphere during one or the other phase of the implosion process. Another series of experiments was based on measurement of the time when the inner surface of an imploded hemisphere made electric contact with one or the other of several 'pins' placed in different positions. A third experiment examined the electric perturbation produced by the implosion of a conductor immersed in a magnetic field. Another experiment used an X-ray flash to photograph a sphere being imploded at various instants of the implosion process. (A drawback of this method was that it could be used only for small gadgets, because of the strong absorption of X-rays in the explosive.) In a fifth experiment, intended to overcome this limitation, plans were made to replace the X-rays with the much more penetrating gamma rays produced by a betatron.

These experiments provided useful information on specific aspects of the implosion process. So did the *terminal* observations, i.e., the examination of objects that had been imploded. But all of this was not enough. Still needed was an experiment capable of studying the implosion of objects of different sizes and, most importantly, capable of following with continuity the development of the implosion rather than catching it at isolated instants of time.

An experiment satisfying these requirements was suggested by Robert Serber in November 1943. The experiment was conceptually simple, but its realization looked difficult and not free of danger. The proposal called for a very strong radioactive source of gamma rays to be placed at the center of the spherical shell whose implosion was to be studied. Fast gamma-ray detectors would be used to measure the radiation emerging from the shell and the surrounding high explosive. The implosion would increase the thickness of the material around the radioactive source, thereby decreasing the intensity of the gamma radiation recorded by the detectors. From this attenuation it would be possible to obtain precise information on the implosion process.

Clearly, the gamma rays should not be too *hard* (otherwise they would not have undergone an appreciable attenuation) or too *soft* (otherwise they would not have been able to penetrate the surrounding material). Keeping in mind these requirements, Serber suggested that the most appropriate radioactive source for the experiment under consideration was *radio-lanthanum*. This is an isotope of lanthanum, with a half-life of about forty hours, which arises from the decay of *radio-barium*, an isotope of barium with a half-life of about 12.5 days. Radio-barium, in turn, is generated abundantly as a fission product, in nuclear reactors. Hence the name of *RaLa experiment* by which the experiment came to be known.

The RaLa experiment required the use of proportional gamma-ray detectors with large sensitive areas needed for the observation of weak gamma-ray fluxes, and capable of faithfully following very fast changes of the radiation intensity. As Serber noted in his proposal, the only known detectors with these properties were the fast ionization chambers then being developed by our group. It was thus natural that we should undertake the task of initiating a program for the study of the implosion by the RaLa method.

We were at an advanced stage in the design of the chambers and auxiliary instrumentation when an unforeseen development threatened to upset all our plans. At the end of April 1944 Luis Alvarez had arrived in Los Alamos from Berkeley. He had been assigned to the Ordnance Division headed by George Kistiakowsky, a scientist from Harvard University, and had been put in charge of a group designated as E-7 or 'RaLa and Electric detonators group'. At the same time, a notice was issued which said, and I quote, 'It will be understood that Mr Alvarez is in general charge of this project (the RaLa project), that Mr Staub of group P-6 will work on the chamber and electronic equipment in cooperation with Mr Rossi.' Staub and I were astonished, to say the least. We did not understand at that time, nor could we ascertain later, what lay behind the decision of shifting the responsibility for the RaLa experiment from ours to another group. At any rate, setting aside personal resentments, we consented to continue our work in the subordinate position we had been assigned.

But as the work went on, it became increasingly clear that with the existing split of responsibilities – one group responsible for the construction of the experimental equipment, and another group charged with the performance of the experiment—the RaLa project was bound to fail. At this point, after discussing the situation with Staub, I called the group together, and explained my view that, as things stood then, the RaLa project was bound to fail, and that it could be rescued only if our group were to take the full responsibility for its execution. I pointed out that this would imply a heavy commitment on the part of all members of the group, who would have to concentrate their efforts on the various aspects of the RaLa program, giving up any other activity. Was the group willing to assume such a responsibility?

The positive response (which I had anticipated) enabled me to secure for our group full control over the RaLa program. We were grateful to Alvarez for his understanding of the situation and for his willingness to accept the change we had requested, while continuing to take care of the many complex logistic problems that the project was facing. Among other things, he was responsible for finding an appropriate location for the experiments. This location was at the bottom of a canyon – the Bayo Canyon – about two miles from Los Alamos. The distance from the inhabited area and the high walls of the canyon made sure that no person unrelated to the project would run any risk from the explosions and the possible spread of the radio-lanthanum. The Alvarez group, no longer engaged in the RaLa experiment, concentrated its efforts on perfecting the electric detonators which held promise of substantially improving the implosion process by ensuring simultaneous detonation of the high explosive at various points of its surface.

At first, our work proceeded without haste because, as I already mentioned, at that time it was not believed that implosion would play a role in the production of the bomb. But the situation changed drastically in the early summer of 1944, when Los Alamos was faced with a problem of such dimensions as to raise the concern that the whole project might come to naught. A short time earlier, a nuclear reactor, the first after the Fermi pile, had started operation in Clinton, near Chicago. By the late spring of 1944 it had produced and shipped to Los Alamos the first gram-size samples of plutonium. Quite unexpectedly, examination of this material had shown that it was a strong source of spontaneous fissions. It was soon ascertained that these fissions came from a contamination of Pu239 by the isotope Pu240, which results from the absorption of a neutron by a nucleus of Pu239. The reason why the presence of Pu240 had not been detected in previous observations on small samples of plutonium prepared with the Berkeley cyclotron was that these samples had not been exposed to the intense neutron flux as the samples prepared with the Clinton reactor had been.

The large number of stray neutrons arising from the spontaneous fission of Pu240 meant that the gun method, on which Los Alamos had relied thus far, could not create a supercritical mass of plutonium sufficiently fast to avoid predetonation. If one was not willing to forego the plutonium bomb whose development was the *raison d'être* of the laboratory, it was necessary to resort to a method of assembly faster than the gun. Implosion was the only method that promised to be capable of producing a sufficiently fast assembly. Therefore it was decided to concentrate all efforts on the development of implosion (although, as I well remember, many scientists at Los Alamos were quite skeptical about the chance of making implosion work in the short time at our disposal). In line with this decision, the laboratory was reorganized. Two new divisions were created. The *Explosive Division* (X-Division) to be headed by Robert Bacher, and the *Weapon Physics Division*, or *Gadget Division* (G-Division) to be headed by George Kistiakowsky. Our group was transferred to the G-Division and renamed G-6, or *RaLa Group*. Alvarez' group was also assigned to the G-Division, and designated G-7 or *Electric Detonator Group*.

Because the RaLa experiment offered the most effective means of diagnosing the performance of the implosion, preparation went ahead with the utmost speed. Soon many other groups became involved. The theorists were suggesting the material and dimensions of the spherical shells, as well as the kind of explosive to be used in order to obtain the most meaningful data about the implosion process. The experts in metallurgy and in explosives had taken upon themselves the task of providing, respectively, the metal shells and the explosive charges. In the canyon, far from the firing site, the radiochemists had set up a laboratory ready to receive the radio-barium and extract the radio-lanthanum from it. It was planned that this radioisotope would then be brought by truck to the firing site enclosed in a container with thick lead walls. Alvarez was in charge of co-ordinating all of these activities.

Our group had set up a 'factory' for the mass production of the ionization chambers. David Nicodemus had the task of designing the chambers, of keeping a check on their construction, and of testing the final product. Benjamin Diven had the delicate job of moving the explosives and the radioactive sources. For this job, he had set up a system of pulleys and strings which allowed him to perform all the necessary operations without coming closer than twelve meters to the radioactive source. Phillipp Grant Koontz (who, as you will remember, had worked with me in Colorado a few years before) was getting ready to analyse and interpret the forthcoming experimental data. Other members of our group had assumed different responsibilities.

The large-area gamma-ray detectors used in the experiment each consisted of several cylindrical fast ionization chambers placed one next to the other and connected in parallel (Plate 34). Four of these detectors were located symmetrically around the gadget, so as to provide information in four different directions of the change in the attenuation of gamma rays during the implosion. They were placed at a sufficient distance from the explosive charge to ensure that they would not be destroyed before recording the signals due to the implosion. These signals were taken by cables to the vertical deflecting plates of four oscillographs, whose horizontal sweeps were triggered simultaneously when the explosive was detonated. The resulting traces on the oscillograph screens were recorded photographically. Examination of the photographs would make it possible to estimate the degree of symmetry of the implosion, the maximum degree of compactness that had been achieved, and the time required to reach this maximum. This was the essential information needed to evaluate the performance of the implosion process.

In the meantime, at the initiative of James Tuck, a member of the British mission, a program had been started aiming at an improved explosive system to be used for the implosion. As already mentioned, ideally the regular implosion of a spherical object would require a spherical, convergent explosive shock wave. At first it was believed that the individual divergent shock waves initiated simultaneously by many detectors uniformly distributed over the surface of a spherical explosive charge would combine into a shock wave closely resembling a spherical convergent wave. Experiments, however, did not entirely bear out this expectation. Tuck's program called for the development of composite units, each formed by layers of explosives of different densities, which would be capable of producing *convergent* shock waves. He argued that several of these units, called *explosive lenses*, arranged around the object to be imploded and fired simultaneously would produce a shock wave closely resembling the ideal spherical convergent shock.

The first shot was fired on September 22, 1944. The object imploded was an iron shell. The RaLa source of gamma rays had a strength equivalent to that of about 40 grams of radium. In this, as in the subsequent shots, each gamma-ray detector consisted of eight ionization chambers. Because of the relatively small strength of the radioactive source, the only result of the experiment was an estimate of the time of collapse of the imploded shell, which turned out to be about 40 microseconds. The second shot was fired on October 4. The arrangement was similar to that of the first shot, except that the imploded shell was made of copper instead of iron and the radioactive RaLa source had a strength equivalent to 130 grams of radium. Because of the greater strength of the source, this shot produced much more significant results than the previous one.

These results, however, were worrisome for they showed that the implosion had been strongly asymmetric.

At this early stage of the experimental program, a young theorist, Robert Christy, came up with a proposal (prompted, primarily by common sense) which was to have an important bearing on the Los Alamos project. This proposal was contained in a short memorandum issued by the Theoretical Division in September, 1944, which read in part:

> Implosion of a Solid Sphere
> Christy has recently made some estimates of the performance of a solid implosion gadget which operates on the compression achieved by a converging shock wave in solid material. The object of this calculation was to find what efficiency could be achieved by an arrangement which probably avoids the difficult problem of obtaining high symmetry and could probably be built with our present knowledge.

As soon as I learned of Christy's proposal, I became anxious to test it experimentally. Therefore, at a meeting held on October 5 for the purpose of deciding the future course of the RaLa program, I insisted that experiments with solid spheres should be included in the program. After some discussion, Oppenheimer agreed with my suggestion and decided that a direct comparison should be made between the implosion of shells and the implosion of solid spheres. In November we started a series of experiments for the purpose of carrying out this comparison. At first, to our disappointment and concern, neither the implosion of shells, nor the implosion of solid spheres gave satisfactory results. Part of the reason might have been that we still did not have explosive lenses at our disposal. But we thought (correctly as it turned out) that the main cause of our troubles was to be sought in the detonators. We were using seventy-two detonators, which were fired by a system of primacords, all starting from the same point. Despite our efforts in ensuring that all branches of the primacord had exactly the same length we never succeeded in firing all the detonators within a sufficiently small time interval.

In January 1945, Alvarez' group finally succeeded in perfecting electric detonators, which could be fired within a fraction of a microsecond of one another. At the beginning of February we carried out two experiments, in which, for the first time, we used the electric detonators. In both experiments the imploded objects were solid cadmium spheres of six centimeters radius (according to the experts in metallurgy, the mechanical properties of cadmium were supposed to closely approximate those of plutonium). The improvement in the performance of the implosion was dramatic. Both shots gave practically identical results. The data from the four ionization detectors showed that the implosion had been almost exactly symmetric and (most importantly) that at the moment of maximum compression the density of the cadmium spheres had increased by a factor of two.

These results, already most encouraging by themselves and which were to be further improved with the use of the explosive lenses, marked a turning point in the program for the development of the plutonium bomb. We continued the experiments with spheres of different materials, with different kinds of explosives, and by making other changes in order to gain a better understanding of the implosion process – a still unfamiliar phenomenon – and to find the optimum performance that could be achieved. We used RaLa sources of great intensity, some greater than one kilogram of radium equivalent.

By the spring of 1945 the plutonium bomb, detonated by implosion (something that less than a year earlier had been little more than dream) had become virtually a reality – a result that, as Oppenheimer was to tell me – could hardly have been achieved in such a short time without the evaluation of the performance of the implosion provided by the RaLa experiment. Soon this 'virtual' reality would become an 'actual' reality in the historic 'Trinity Test' when, as a fulfillment of the activities of the Los Alamos scientists and as a test of their results, the phenomenon of the nuclear explosion was submitted to a direct experimental check.

But before leaving the subject of our work in Bayo Canyon I wish to give a chronological account of a typical experiment, as a concrete example of the procedure we actually followed in our work.

Shot No. 25 of June 8, 1945
The model of the bomb is a solid cadmium sphere, surrounded by explosive lenses to be fired by electric detonators. An aluminum

shell inserted between the cadmium sphere and the high explosive is used to equalize the pressure exerted over the cadmium surface by the explosive shock. A hole is drilled, reaching from the outer surface of the aluminum shell to the center of the cadmium sphere. This hole will be used to insert the radioactive source.

June 7, pm
The gadget is mounted on a light wooden frame, with the hole in the cadmium sphere and in the aluminum shell pointing in the vertical direction.

June 8, am
Machinery is installed to be used for performing, by remote control, all operations involving the risk of exposure to radiation. The detonators are attached to the explosive lenses and are connected by cables to a command circuit located in a bomb-proof shelter. Four trays of ionization chambers are mounted at a proper distance, around the explosive charge. The chambers are cylindrical in shape, 2 inches in diameter and 30 inches long, with a wire stretched along the axis. They are filled, at 4.5 atmospheres, with a mixture of argon and carbon dioxide. The wires of the eight chambers in each tray are connected to one another and brought to ground through a resistor and a galvanometer, both located in the shelter. After going through a preamplifier and an amplifier, the signal from each tray of ionization chambers (as represented by the potential drop along the resistor) is taken to the vertical deflecting plates of a separate oscillograph. The horizontal deflecting plates of the four oscillographs are connected together and to the output of a circuit which, on command, produces in each oscillograph an horizontal sweep of the electron beam. Preparations are made to photograph the traces that will appear on the screens of the oscillographs when the shot is fired.

June 8, pm
A number of control experiments and calibrations are carried out. In order to obtain an absolute measurement of the current in the ionization chambers, a known electric current is sent through each of the four ground resistors, and the corresponding deflections of the electron beams in the four spectrographs are measured. The

speed of the horizontal sweeps is measured with a suitable instrument. In the meantime our colleagues in the radiochemistry laboratory have prepared a RaLa source equivalent to about 850 g. of radium, and have placed it at the tip of a rod, which will be inserted in the gadget.

From 15:00 to 15:30 the rod enclosed in a thick-walled lead container is taken by a truck to the firing site (Plate 40). By remote control, a section of the explosive above the hole is lifted. Still by remote control, the rod carrying the radioactive source is taken out of the lead container and inserted into the hole so as to place the source at the center of the cadmium sphere. The rod itself fits into the hole, so as to fill the gaps in the cadmium and aluminum spheres. The section of the explosive which had been removed is now replaced, and the gadget is ready for the test.

At 16:49, with everybody in the shelter, the shutters of the photographic cameras are opened and, immediately after, the explosive lenses are detonated. An electric signal produced by the explosion starts the electron beams in the four oscillographs and, at the same time, it activates the electronic circuit which starts the horizontal sweeps. The experiment is now completed, and it only remains to develop the photographic films. The traces that appear on these films show initially, as expected, a fast decrease of the gamma-ray intensity due to the compression of cadmium core. After going through a minimum the gamma-ray intensity begins to rise again; this rise is due to a decompression and eventually to a dispersion of the material surrounding the source (see Fig. 4.1).

Notice the exceedingly short duration of the period of maximum compression. In order to obtain the maximum energy release from the nuclear explosion the chain reaction must be initiated during this period. Initiators activating neutron sources at the proper time were under development elsewhere in the laboratory. RaLa experiments, such as the one described, determine the accuracy that is required in the timing of these initiators.

Finally, a note about the question of safety for the people working in Bayo Canyon. As we were preparing the experiment, we had some concern about potential risks to the personnel obliged to handle powerful high explosives and radioactive sources of unusual strength. Actually, thanks to the meticulous precautions that we

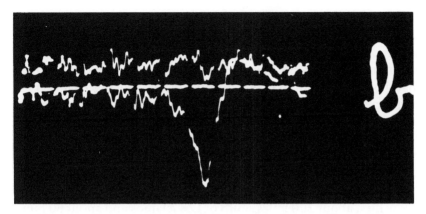

Fig. 4.1. Photographic record of one of the four oscillograph traces obtained in the RaLa shot of June 8, 1945. The initial fast decrease of the gamma-ray intensity is due to the compression of the cadmium sphere. The subsequent increase is due to a decompression followed by a dispersion of the material around the gamma-ray source. (Los Alamos Files)

had taken, we never had to worry in Bayo Canyon about dangerous situations arising from premature explosions or by the exposure to radiation. If there was a risk in our operation, it was to be found in the transfer of personnel between Los Alamos and Bayo Canyon. The roads were in very poor condition, and the cars were army rejects. The combination of these two factors was responsible for an accident which almost had serious consequences. This happened when Morton Levin, one of our SED (Special Engineering Detachment, a specialized corps of the army), while driving from Los Alamos to the Bayo Canyon, lost control of the car which fell into the canyon below. Fortunately the steel roof of the car withstood the impact, and Morton escaped with a number of bruises and a severe psychological shock.

By attempting to reconstruct the incident, we arrived at the following likely explanation. Because of the recent rains, the dirt road had become like a washboard marked by small nearly equidistant transverse ridges. At a certain car speed the periodic impulses applied to the car by these ridges would come into resonance with the proper vibration of the car itself. At this point the car would start to oscillate violently up and down until it went out of control. Inadvertently, Morton must have reached this critical speed.

I now come to the culminating phase of the Los Alamos project, to the experiment that was intended as the final crowning of our efforts. Preparations for this experiment had already begun in the autumn of 1944 in the hope that the tests on the implosion, which had just begun, would give a positive result. The place chosen for the experiment was an area of about 400 square miles, part of the military base of Alamogordo, in the region of the desert of southern New Mexico which had been first explored by the Spaniards and which, from them, had received the (prophetic!) name of *jornada del muerto*. This location, which had been given the code name of *Trinity* – a name evoking other secrets, other mysteries – was, as the crow flies, about 200 miles south of Los Alamos and about 60 miles north of the Mexican border.

Preparations had been made for a number of observations to be done during the explosion in order to evaluate the performance of the bomb. In the first place, of course, it was necessary to estimate the total energy released in the explosion. This evaluation was the objective of several experiments based on the observation of one or the other effect of the nuclear explosion (such as the Earth shock, the air blast, the neutron emission, the gamma-ray emission, etc). Of these experiments I shall report just one, performed by Fermi, that is a typical example of his unique ability to obtain important results with the simplest means. But I let Fermi speak:

On the morning of the 16th of July, I was stationed at the Base Camp at Trinity in a position about ten miles from the site of the explosion.

The explosion took place at about 5.30 A.M. I had my face protected by a large board in which a piece of dark welding glass had been inserted. My first impression of the explosion was the very intense flash of light, and a sensation of heat on the parts of my body that were exposed. Although I did not look directly towards the object, I had the impression that suddenly the countryside became brighter than in full daylight. I sub-sequently looked in the direction of the explosion through the dark glass and could see something that looked like a con-glomeration of flames that promptly started rising. After a few seconds the rising flames lost their brightness and appeared as a huge pillar of smoke with an expanded head like a gigantic mushroom that rose rapidly beyond the clouds probably to a height of the order of 30,000 feet. After reaching its full height, the smoke stayed stationary for a while before the wind started dispersing it.

About 40 seconds after the explosion the air blast reached me. I tried to estimate its strength by dropping from about six feet small pieces of paper before, during and after the passage of the blast wave. Since, at the time, there was no wind I could observe very distinctly and actually measure the displacement of the pieces of paper that were in the process of falling while the blast was passing. The shift was about $2\frac{1}{2}$ meters, which, at the time, I estimated to correspond to the blast that would be produced by ten thousand tons of T.N.T.

I only wish to add that the result obtained by Fermi differs by less than a factor of two from the average of all the results obtained by other scientists with much more complex experiments.

Another quantity of basic importance for the evaluation of the performance of the bomb was the rate of increase of nuclear activity, because, as I have already explained, the energy released by a chain reaction depends on the relation between this quantity and the rate of thermal expansion of the fissionable material. Oppenheimer had placed Robert Wilson in general charge of a research program directed at measuring the rate of growth of nuclear activity. As part of this program, I proposed an experiment which, as it turned out, was the only fully successful one of three experiments attempted.

I was well aware of the difficulties that I would be facing in performing this experiment—difficulties due primarily to the exceedingly short timescale of the phenomenon, on the order of 10^{-8} seconds. (In 1945, observations with timescale on the order of 10^{-8} seconds were almost beyond the state of the art.) I wish to make clear that the final success of the experiment was due, to a large extent, to the co-operation of a number of people from my own group as well as from other groups. Among these, was Herbert Bridge, who is still a close colleague and friend.

The experiment was designed to measure the rate of increase of gamma radiation emitted during the chain reaction, the intensity of which was known to be proportional to the rate of the nuclear activity. It was expected that this increase would follow an exponential law of the type: $e^{\alpha t}$. Therefore the result of the measurements could be described by means of the coefficient α. Hence the name of 'α-experiment' by which the experiment became known. Note that the quantity $1/\alpha$ represents the time during which the gamma-ray intensity increases by a factor of e, a time which, as I already mentioned, was expected to be on the order of 10^{-8} seconds.

The first task was the choice of a suitable gamma-ray detector, a task made harder by the fact that it would not have been possible to check its performance beforehand under conditions approaching those in which it was going to be used. To my knowledge, the only detector whose properties came close to those needed for the α-experiment was the fast ionization chamber. But the α-experiment required of the detector a response much faster than had been sufficient in previous experiments, such as the RaLa experiment. Therefore I had to re-examine the operation of the ionization chamber, to see whether it was possible to further increase its speed of response. I found that, indeed, I could produce a detector of the required speed by means of an ionization chamber of rather unusual design, whose sensitive volume was a narrow space between two coaxial cylinders. In the detector that we built for the α-experiment the outer cylinder was seven inches in diameter, five feet in length. The spacing between the outer and inner cylinder measured $\frac{3}{8}$ inches. It was filled with the usual mixture of argon and carbon dioxide. Even without the benefit of an experimental test, I felt confident that an ionization chamber of this particular design would be able to closely follow the very fast variation that was expected of the gamma-ray intensity – as, in fact, it did.

The second problem was the choice of recording system. Since we had to deal with an experiment which could not be repeated, reliability was the first requirement. We had to use a system as simple and as direct as possible, that would offer the highest guarantee against malfunction and against any ambiguity in the interpretation of the experimental data. With these requirements in mind, we discarded all complex systems of electronic recording, and decided to represent the signal of the ionization chamber graphically, as a trace on the screen of an oscillograph. Oscillographs sufficiently fast for our purpose were not yet available commercially. However we learned that at the Dupont Company attempts were being made to develop faster oscillographs. Thus we contacted this company and received three experimental oscillographs, of which one was just barely fast enough for our experiment.

It still remained for us to decide exactly in what manner the signal from the oscillograph should be translated into a mark on the trace on the screen. The usual method for recording an electric signal was (and still is) to apply the signal to the plates that control the vertical displacement of the electron beam, while at the same time applying a saw-tooth signal to the horizontal plates by an electronic circuit, thereby producing a horizontal sweep of this beam. Clearly, this system could not be used in our case, because it was practically impossible to start the horizontal sweep exactly at the proper time – and *exactly* here means a few hundreds of a microsecond from the beginning of the chain reaction.

In order to overcome this difficulty, I thought of a new system, which consisted of applying the signal to be recorded to the *horizontal* deflecting plates of an oscillograph, and applying the sinusoidal voltage output of a quartz high-frequency oscillator to the *vertical* deflecting plates. The trace produced by an exponential signal would have the shape of a 'stretched out' sinusoid with successive 'wave lengths' increasing in geometric proportion, as shown by the simulation in Fig. 4.2. From the ratio between two successive 'wave lengths' it would be possible to obtain an accurate value for the rate of increase of the signal.

Since the ionization chamber must be placed near the bomb, while the recording instrument must be located at a safe distance, it was necessary to use a transmission line to bring the output signal of the former to the latter. To design such a line was not a trivial problem. The line was required to transmit the signal without any distortion or attenuation. In fact, it was desirable that the line should amplify the signal because I did not trust the electronic amplifiers existing at that time.

I looked for some treatise that would help solve my problem, but did not find what I needed. I then decided to go back to fundamentals. Considering a line made of separate sections, I made the

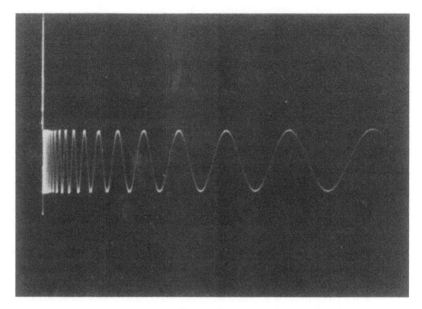

Fig. 4.2. Trace on an oscillograph screen of a simulated exponential signal recorded with the method described in the text.

condition that, in each section, the electromagnetic field should obey Maxwell's equations; I then applied the conditions of continuity to the points of contact between neighboring sections.

In this manner, I succeeded in computing the structure of a transmission line suited to our experiment. However, since the transmission line was such an essential part of our instrumentation, I felt that, in order to be doubly certain of my results, I should verify them with someone more expert than I on the matter of transmission lines. I. I. Rabi, who was in Los Alamos on one of his frequent visits, recommended Edward Purcell as the scientist most knowledgeable in this field. At that time, Ed was working on problems of radar at the Radiation Laboratory of the Massachusetts Institute of Technology in Cambridge, Massachusetts. At my suggestion he was invited to Los Alamos as a consultant. I met him on his arrival at Albuquerque airport, and, during the trip by car from Albuquerque to Los Alamos, I brought him up to date on our work on the hill. He was flabbergasted. He did not have the faintest idea that we were trying to develop a nuclear bomb, and that, in fact, such a bomb was about to be exploded.

At first there was some problem of communication between me, speaking the language of first principles, and Ed, an expert in the applications of science, speaking the language of the specialist. But these difficulties were soon ironed out; Ed approved my results and made some useful additions (which, if I remember correctly, had to do with the terminations of the line).

The transmission line, built according to the design derived from our study, was essentially an oversize coaxial cable. The outer wall was a copper pipe, three inches in diameter. The inner conductor consisted of several sections whose diameter decreased stepwise from the ionization chamber to the oscillograph. The diameters of the first and last sections were chosen so as to match the low impedance of the ionization chamber at the entrance, and the high impedance of the oscillograph at the exit. The change of impedance along the line was expected to produce the desired enhancement of the signal from the ionization chamber.

Having prepared the instrumentation in Los Alamos a few weeks before the expected date of the test, I went down to Trinity with my collaborators to set up the experiment. After the bomb had been installed in its place, on top of a steel tower 100 feet high, we mounted the ionization chamber at a suitable distance from it. We attached the transmission line to the chamber and brought it down at an angle to a point three feet underground, 200 feet from the tower. From there the transmission line continued, still underground, to an underground room, where we installed the oscillograph and the auxiliary instruments. (This particular path of the transmission line had been chosen so as to avoid the danger of interference by the gamma radiation produced during the explosion.)

We spent weeks doing tests and calibrations over and over again, in order to make sure that the experiment would work properly and that we would be able to interpret the records. It was a period of hard work, in the intense heat of the desert, of strong anxieties, of exhausted waiting. Finally, the fatal day arrived. On July 16, 1945 at 5.30 in the morning the first nuclear bomb was detonated. I, with a few colleagues, was watching from a place some twenty miles

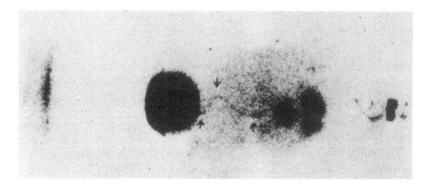

Fig. 4.3. An historical picture. From the faint trace indicated by
the arrows it was possible to measure the initial rate of increase
of the nuclear activity of the bomb exploded in the Trinity test.
(Los Alamos Files)

from the tower. Others have described the blinding flash, the
deafening thunder, the gigantic mushroom cloud; there is no need
for me to repeat.

Shortly after the explosion I started for Los Alamos, a trip of
some 350 miles; with me in the car were Benjamin Diven, Otto
Frisch, and a WAAC. Otto offered to share the driving, but I did not
dare accept because he was the worst driver I had ever known. One
after the other my passengers fell asleep, and I was left alone with
my thoughts. Until then the pressure of the work had been such as to
leave no time for reflections. Now the terrifying significance of what
we had done hit me like a blast. I must admit that at times I felt a
certain pride at having played a role in an undertaking of such great
difficulty, of such historical importance. But soon this feeling was
overwhelmed by a feeling of guilt and by a terrible anxiety for the
possible consequences of our work, a guilt feeling that would be
reinforced a short time later when I learned of the destruction of
Hiroshima and Nagasaki. Like many of my colleagues, I had hoped
that the bomb would be used in a bloodless demonstration to
induce Japan to surrender. I arrived in Los Alamos exhausted, and
slept for an entire day and night.

A few days later, when the radioactivity of the ground at the test
site had sufficiently decayed, my instruments were recovered and
brought to Los Alamos. I took the film from the camera, and

brought it to the darkroom. Fermi was with me, and we were anxiously watching as the film was developed. Finally a trace appeared, of the predicted sinusoidal shape, very faint, but perfectly visible: it is reproduced in Fig. 4.3. Even before the film was dry, it was possible to compute with high precision the initial rate of growth of the nuclear activity.

We stayed in Los Alamos a few more months. There were still a few things I had to take care of. Also, Staub and I were still working on a book, *Ionization Chambers and Counters*, based primarily on our work on fast ionization chambers. By then our family had grown by the arrival of Frank, our second child, born in the Los Alamos Army hospital one year earlier. But we were in a hurry to leave, which we did on February 6, 1946, with a mixed feeling of relief and nostalgia.

5

Cosmic rays at MIT
(1946–   )

Spring of 1946

I was now at the Massachusetts Institute of Technology (MIT) as a professor in the Physics Department, a position that had been offered to me toward the end of my stay in Los Alamos. Eager to resume my usual activity, after the digression of Los Alamos, I was preparing to teach and was planning a research program. As part of my educational activity, I had begun writing a set of textbooks: *High-Energy Particles* (to be published in 1952), then *Optics* (1957), *Cosmic Rays* (1964), and *Introduction to the Physics of Space* (with Stanislaw Olbert in 1970).

Four young physicists had come with me from Los Alamos: Herbert Bridge, Mathew Sands, Robert Thompson, and Robert Williams. They formed the core of the group which soon grew, as other young scientists became interested in joining it. The first to do so were John Tinlot and Robert Hulsizer, who, during the war, had worked on the development of radar at the Radiation Laboratory of MIT. Like their four colleagues from Los Alamos, they had come to MIT as Ph.D. candidates. But, because of their experience in war work, all six were much better prepared for independent research activity than the average graduate student. Most welcome was the support I received from a senior physicist, George Valley, in my efforts to get our research program under way.

From the outset, and throughout its life, our group attracted scientists from other institutions in the United States and from many foreign countries (France, Italy, China, Japan, India, Australia, Brazil, Bolivia, etc.). By their presence, our foreign guests lent the group the quality of an open international society. Their contribution to our research program was largely responsible for the renown acquired by our group in the scientific community.

International interaction extended beyond the boundaries of the group. As an outstanding example, I recall the frequent visits of Vikram Sarabbai, a distinguished Indian physicist, with whom we had developed strong ties of friendship and of scientific interest. Vikram was a man with a warm, endearing personality. He belonged to an influential family which was largely responsible for the growth of the textile and pharmaceutical industries in India. In addition to his activity as a physicist, Vikram was deeply involved in the political, economic, and cultural life of his country. Many of his friends suspect that overwork was the cause of his early death.

Ultimately, almost all of our foreign guests returned home where, helped by their experience at MIT, they made substantial contributions to the scientific and educational life of their countries. Some were called to fill positions of great responsibility and prestige. For example, C. Y. Chao eventually became the Director of the Institute for Nuclear Physics in Peking. Bernard Gregory was for several years the Director of CERN/Centre Européen de Recherches Nucléeres, the most important international center of nuclear physics, and was then appointed Director of the French National Research Center. Minoru Oda was instrumental in creating, in Japan, one of the world's most important centers of space science and technology and is now President of the Japanese Research Institute for Physical and Chemical Research. Yash Pal is chairman of the Grant Commission of the Indian Government. B. V. Sreekantan is Director of the Tata Institute of Fundamental Research in Bombay.

I was well aware that in my new position, my activity would be very different from what it had been in past years. Then, working alone or, at most, with the help of a few students, I would build the instruments needed for my experiments, I would take them to the place where they had to be used, I would make the measurements, and analyse the results. Now I had the responsibility of an entire group, and what mattered was no longer my own work but the work of the group. In the first place my task was to identify the most promising research programs among those that were within our reach. I had then to help where help was needed in the planning of the instrumentation and evaluation of the experimental results, all of this without discouraging the individual initiative of the researchers.

In the years that followed the end of the war, the renewed interest in the fundamental aspects of science had acted as an incentive to resume the study of cosmic rays, a study which, for about three years, had remained practically dormant. Two sets of basic problems still required close examination: (1) problems regarding the primary radiation, and (2) problems regarding the nuclear interactions of cosmic rays and the particles generated in these interactions. It was on these two problems that our group chose to concentrate its efforts. But, before I discuss our program, I shall briefly summarize the information that was available at the start of our work.

Primary cosmic rays

The East–West effect had shown that the primary cosmic radiation consists, for the most part at least, of positively-charged particles. It was generally assumed that these particles are protons but one could not rule out the possibility that a small number of electrons and/or photons might also be present. The only firm piece of evidence about the energy spectrum of the primary radiation came from the observation of a substantial flux of cosmic rays in the equatorial region of the Earth. This meant that the spectrum extended well beyond ten billion electron volts (10^{10} eV), this being, according to theory, the minimum energy of singly-charged particles capable of reaching the geomagnetic equator.

Nuclear interactions

Nuclear interactions of cosmic rays were first observed in the early 1930s by Marietta Blau and H. Wambacher. In photographic emulsions, exposed for a long period of time to cosmic rays, they discovered a number of so-called *stars*, i.e., groups of heavily ionizing particles diverging from one point, which they correctly identified as the fragments of atomic nuclei disrupted by the impact of cosmic-ray particles, a phenomenon classified as a *low-energy nuclear interaction* because it did not require a large expenditure of energy.

High-energy nuclear interactions were discovered some ten years later by Gleb Wataghin in Brazil and Ludwig Jánossy in England. Using arrays of G.M. counters, separated by layers of lead, they observed a number of coincidences which could have been produced only by groups of associated penetrating particles. Wataghin and Jánossy interpreted the occurrence of these groups, to become known as *penetrating showers*, as evidence for the multiple production of mesotrons in nuclear interactions of cosmic rays, a process described as a *high-energy nuclear interaction* because it required the expenditure of a large amount of energy. Penetrating showers, however, had not yet been studied in much detail, nor was it known by which component of the local cosmic radiation they were produced.

Elementary particles

Until 1946, in addition to the particles of which atoms are made (protons, neutrons, negative electrons) the only known elementary particles were the positive electron and the mesotron (and perhaps the neutrino). As I mentioned repeatedly above, mesotrons are unstable particles whose mean life was exactly known from the experiment of Nereson and myself at Cornell. It was also known that electrons were among their decay products. Conservation of momentum required that the emission of electrons be accompanied by the emission of neutral particles, presumably neutrinos. Still debated was the question of whether one or several neutrinos were produced in each decay process.

This, in brief, was the state of our knowledge in 1946. But in the physics of elementary particles a revolution was about to break out. In 1947, shortly after an experiment by the Italian physicists Marcello Conversi, Ettore Pancini, and Oreste Piccioni had shown that the mesotron could not be the particle associated with the nuclear field of force, a group of physicists in Bristol including Cesare Lattes, Giuseppe Occhialini, and Cecil Powell, announced an important discovery. Using the newly developed nuclear emulsion, they observed a previously unknown elementary particle which was generated in the nuclear interactions of cosmic rays and promptly decayed giving rise to a mesotron. This particle was named *π-meson*, or *pion*. The mesotron was rebaptized *μ-meson* or *muon*.

Shortly thereafter, George Rochester and Clifford Butler in Manchester published cloud chamber photographs of two unusual events. The first photograph suggested the decay in flight of a neutral particle into two charged particles. The second photograph suggested the decay in flight of a charged particle into a second charged particle and, supposedly, a neutral particle. Surprisingly, neither the neutral particle needed to explain the event seen in the first picture, nor the charged particle seen to decay in the second picture could be identified with any previously known particle. These particles were therefore the first examples of a new kind of object, and were temporarily described as *new particles*.

The activities of our group were essentially of an experimental nature. Therefore, technological facilities played an essential role in our research program. G.M. counters and cloud chambers were widely used as in previous cosmic-ray work. To these familiar instruments two new devices were added: the *fast ionization chamber* and the *scintillation counter*. Scintillation counters suitable for cosmic-ray research were developed, some time later, by our group, as the basic tools for one of its experiments. I shall describe them in conjunction with this experiment. Here I wish to add a few words about the fast ionization chamber.

As explained in the previous chapter, the fast ionization chamber had been developed in Los Alamos for the purpose of measuring very rapid variations of an ionizing radiation. It consists of a cylindrical metal container filled with a gas and with a metal wire stretched along the axis. In cosmic-ray research it can be used to detect bursts of ionization in the gas produced by one of several events associated with cosmic rays, such as groups of heavily ionizing particles ('stars') arising from the disintegration of atomic nuclei, electronic showers generated locally by high-energy electrons or photons, penetrating showers, and extensive air showers. With the wire brought to a positive potential through a resistor, the chamber will respond to the production of a burst of ionization in the gas with a short electrical pulse at the wire, of a size proportional to the number of ion pairs in the burst.

Herbert Bridge and others in the group developed methods for identifying which of the events listed above were mainly responsible for the ionization bursts recorded by a fast chamber operated in different conditions and at different altitudes above sea level. These methods included operating the ionization chamber inside a cloud chamber, analysing the shape of the electric pulses (which reflects the distribution of the ionization in the volume of the chamber), and selecting bursts which occur simultaneously in several fast ionization chambers and therefore must be due to extensive air showers, etc. But to describe in more detail this important phase of our program would take me too far afield, so I shall turn from technology to science.

The years which I spent with the MIT cosmic-ray group, participating directly or indirectly in its work, were an integral part of my life as a scientist. Thus it would please me to describe in some detail all of the manifold research activities of this group during those years. Since this description would exceed the limits I have set to the length and the degree of technical complexity of this story, I shall confine my remarks to some of the most significant accomplishments.

One of the first problems addressed by the group concerned the nature of the radiation responsible for the nuclear interactions of cosmic rays. For many years it was known that the local cosmic radiation consists essentially of electrons, photons, and μ-mesons. Both theory and experiment had shown conclusively that one could not identify the nuclear-active particles with electrons or photons. Thus it was necessary to conclude that the nuclear interactions were produced either by μ-mesons, or by some rare component of the local radiation which, until then, had escaped detection. (At that time the fact that μ-mesons do not respond to nuclear forces had not been clearly established.)

To decide between these two possibilities John Tinlot, between 1947 and 1948, undertook to measure the variation with altitude of the penetrating showers produced in a lead block. From a comparison between the results of these measurements and the known altitude dependence of the μ-meson intensity it should have been possible to decide whether or not μ-mesons might be responsible for the production of penetrating showers.

The instrument used by Tinlot was a penetrating shower detector of the type employed by Wataghin and Jánossy. Measurements taken at sea level, at different heights in the Rocky Mountains of Colorado and aboard a B-29 aircraft, up to a maximum height of

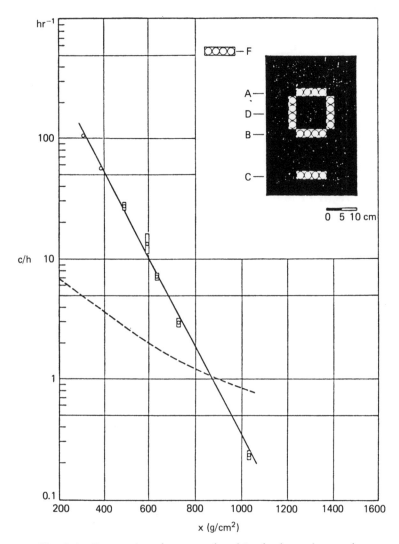

Fig. 5.1. Comparison between the altitude-dependence of penetrating showers (solid curve) and the altitude-dependence of the μ-mesotron intensity (dotted curve). The abscissa is the mass of the air layer above the instrument in g/cm²; the ordinate is the intensity on a logarithm scale. The insert shows the penetrating-shower detector. From an article by J. Tinlot in *Physical Review*, **73**, 476 (1948).

30 000 feet, showed that the frequency of penetrating showers increases with altitude much more rapidly than the intensity of the μ-meson component (see Fig. 5.1). For example, from sea level to the top of Mt Evans, at an altitude of 14 300 feet, the μ-meson intensity increases by a factor of 2.5, whereas the frequency of penetrating showers increases by a factor of 32.

Hence the rays responsible for the nuclear interactions could not be identified with μ-mesons. (As we shall see later, the nuclear interactions are produced by protons, neutrons and, occasionally, by π-mesons.)

The above conclusions apply to the radiation responsible for the high-energy interactions which manifest themselves with the production of penetrating showers. An experiment performed by Bridge showed that they also apply to the radiation responsible for the low-energy nuclear interactions. This result was obtained from a measurement of the altitude dependence of the bursts in an unshielded pulse chamber; previous tests having shown that, in the absence of any shield, almost all bursts recorded by a fast ionization chamber are produced by cosmic-ray 'stars'.

I wish to add here a few words about another early experiment initiated by Robert Thompson. Regretfully, this experiment was not brought to a completion, but is still worthy of notice as an example of a refined experimental procedure. The experiment was intended to answer the question concerning the number of neutrinos produced by the decay of μ-mesons. If only one neutrino was produced, then all decay electrons would have the same energy. If two or more neutrinos were produced, then the electrons arising from different decay processes would have different energies.

The instrument used by Thompson to measure the energy of the decay electrons was a cloud chamber containing a metal plate and operated in a magnetic field. The decay electrons of μ-mesons stopping in the plate would emerge into the sensitive volume of the chambers, where their energy could be measured by the curvature of their tracks in the magnetic field. One of the cloud chamber pictures obtained by Thompson is reproduced in Fig. 5.2. The experiment was interrupted by Thompson being called to the University of Indiana, and others would get the credit for answering the question addressed by him. (It was found that each decay process of μ-mesons produced two neutrinos.)

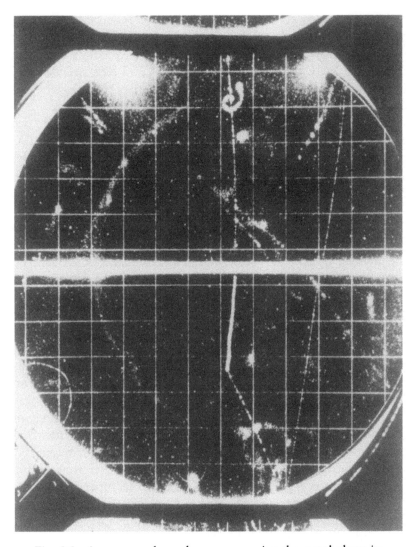

Fig. 5.2. A μ-meson slows down on crossing the metal plates in
the cloud chamber. It then decays producing an electron. From
an article by R. Thompson, in *Physical Review*, **74**, 490 (1948).

I now turn to a particularly important and exacting research
program, aimed at broadening our knowledge of nuclear interactions of cosmic rays, which primarily required a detailed study of
the products of these interactions. Most of our group was engaged

in this program; major contributions were made by Herbert Bridge, Bernard Gregory, Charles Peyrou, Wayne Hazen, John Tinlot, and Martin Annis.

Many of our experiments were performed by means of large rectangular cloud chambers outfitted with several horizontal plates of different materials and different thicknesses (see Fig. 5.3). These multiplate chambers proved to be very useful for the observation of events that could not be detected in photographic emulsions or with cloud chambers of a different design. We operated our chambers sometimes at sea level, but more often in the Rocky Mountains of Colorado, so as to take advantage of the fast increase of the frequency of nuclear interactions with altitude.

Thus I had an opportunity to revisit the places which I had known in the first years of my stay in the United States. I spent the summer of 1949 with my family, on a ranch, a short distance from Echo Lake. The paved road now reached Summit Lake. MIT had built, near Echo lake, two log cabins, one to be used as a dormitory, the other as a laboratory. Otherwise there were few changes. The lake was still surrounded by a dense forest of spruce, into which one could not venture without running the risk of getting lost. In fact, this happened to one little girl, the daughter of one of our colleagues, who, fortunately, was later found, safe and sound. Wildlife was still plentiful: squirrels and *camp-robbers*, birds which preyed on our supplies, bears which came at night to rummage in our garbage. And every morning a fawn would come out of the forest to play with the children.

I shall now summarize briefly the information about nuclear interactions and elementary particles which is contained in our cloud-chamber pictures (see examples in Figs 5.3–5.8). In the first place, we found that nuclear interactions are produced both by ionizing (and therefore electrically-charged particles), and by non-ionizing (and therefore neutral) particles. The latter are certainly neutrons. The small density of the tracks of the charged particles indicates that these have the low ionizing power which characterizes singly-charged particles moving with a velocity close to that of light. Undoubtedly most of these particles are high-energy protons. Some, presumably, are π-mesons generated in nuclear interactions above the sensitive volume of the chamber.

Among the products of nuclear interactions one can often recog-

Fig. 5.3. The cloud chamber contains 0.5-inch thick brass plates. A nuclear interaction in the uppermost plate produces a large electronic shower and a few penetrating particles. (MIT Cosmic-ray Group)

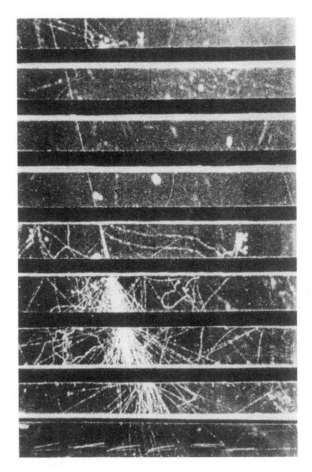

Fig. 5.4. The cloud chamber contains 0.5-inch thick brass plates. An ionizing particle, probably a proton, traverses several plates undisturbed, then undergoes a nuclear interaction producing a mixed shower. (MIT Cosmic-ray Group)

nize both electronic showers and penetrating particles which traverse several of the plates in the multiplate cloud chamber. Sometimes the penetrating particles appear in groups which, undoubtedly, are to be identified with the penetrating showers that had been observed previously by coincidence experiments with G.M. counters.

Efforts were made to discover the nature of the penetrating

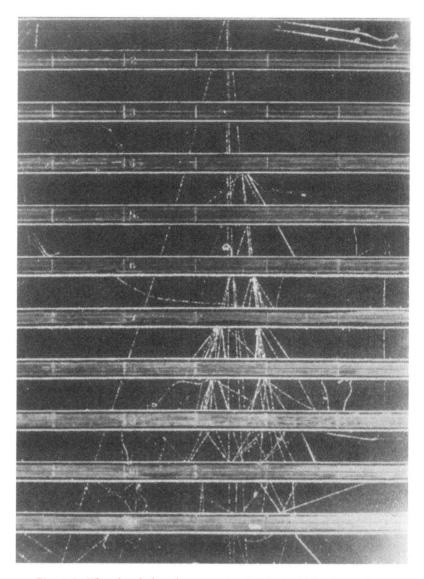

Fig. 5.5. The cloud chamber contains 0.5-inch thick plates. A penetrating shower is produced in the uppermost plate. One of the penetrating particles undergoes a nuclear interaction in the fourth plate. (MIT Cosmic-ray Group)

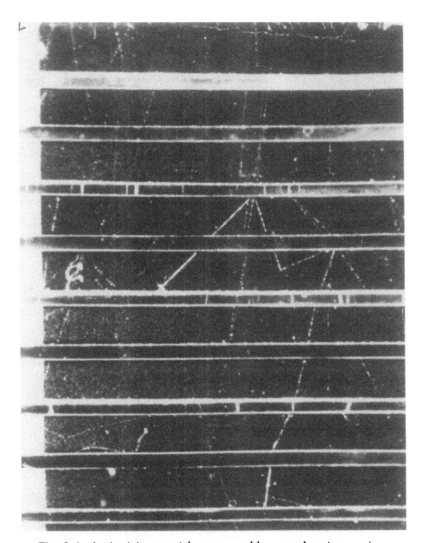

Fig. 5.6. An ionizing particle generated by a nuclear interaction in a lead plate traverses an aluminium plate and halfway between this and the next plate undergoes a decay process ejecting a meson which is scattered downward in the aluminum plate. From an article by H. Bridge & M. Annis in *Physical Review*, **82**, 445 (1951).

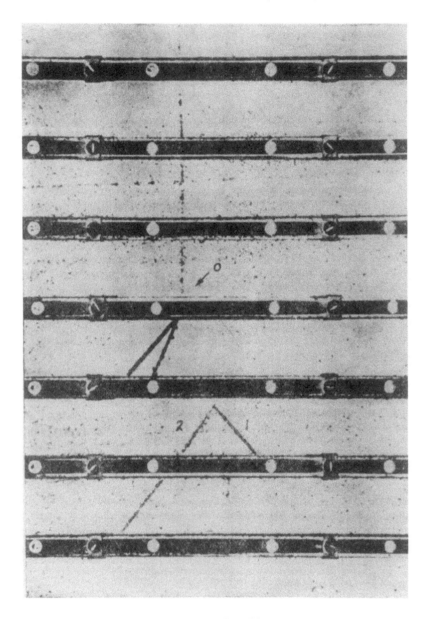

Fig. 5.7. A neutral hyperon produced by a nuclear interaction
in a lead plate decays producing two ionizing particles. The one
to the right is a proton; the one to the left, a meson. From an
article by H. Bridge, C. Peyrou, B. Rossi & R. Safford, in
Physical Review, **91**, 362 (1953).

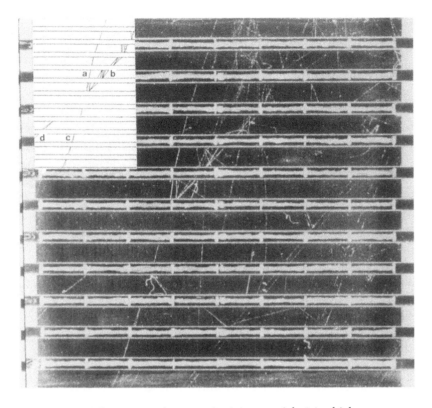

Fig. 5.8. The picture shows an ionizing particle (*a*) which crosses four plates and stops in the fifth plate, giving rise to an electronic shower (*b*) moving upward and two gamma rays moving downward which also initiate electronic showers (*c*) and (*d*). The total energy set free in the production of these secondary rays is greater than 1470 ev. After ruling out other possible interpretations, we concluded that the incident particle is probably a negative proton or a negative hyperon which undergoes an annihilation process by colliding with a proton. From an article by H. Bridge, H. Courant, H. De Staebler & B. Rossi, in *Physical Review*, **95**, 1101 (1954).

particles; the first step was an attempt to estimate their mass. The idea behind this was that, just as one could compute the *observable* properties of a particle (ionizing power, scattering, range) from a knowledge of its *physical* properties (mass, velocity, etc.) so it must be possible to estimate the physical properties of a particle, and, in particular its mass, from the observation of its cloud-chamber tracks (density of the track, scattering in the traversal of the plates, range of a particle stopping in one of the plates). Using a procedure developed by Stanislaw Olbert on the basis of this idea, we found that the mass of the penetrating particles was close to the meson mass. However the accuracy of the mass determination was not sufficient to distinguish between π-mesons and μ-mesons. The uncertainty was removed by some photographs showing penetrating particles which suffered a nuclear interaction in one of the plates (Fig. 5.4). Since, by that time, it had been shown conclusively that μ-mesons do not interact with nuclei, we could safely conclude that the penetrating particles born from nuclear interactions are π-mesons.

The electronic showers that are observed in the cloud-chamber pictures of nuclear interactions show that high-energy electrons or photons are often products of these interactions. A hypothesis advanced by Oppenheimer in 1947 offered the best explanation of these observations. He pointed out that the production of charged π-mesons should be accompanied by the production of *neutral* π-mesons, and that these should decay promptly, each producing two photons.

If neutral π-mesons were indeed responsible for the production of electronic showers, one had to anticipate that these showers are initiated by photons rather than by electrons. This means that there should be a non-ionizing link – a *gap* – between the point where the nuclear interaction occurs and the point of origin of the electronic shower. Of course it would be possible to verify the existence of a non-ionizing link only if the process which initiates the electronic shower (pair production by the photon) does not occur in the same plate where the nuclear interaction takes place. Guided by these considerations, Gregory and Tinlot took a number of pictures with a cloud chamber containing alternately lead plates $\frac{1}{4}$ inches thick, and aluminum plates $\frac{5}{16}$ inches thick. Both from experiment and theory it was known that photons have a negligible probability of initiating an electronic shower in a thin plate of a substance of low atomic number such as aluminum. And, in fact, of all the electronic showers associated with nuclear interactions occurring in one of the aluminum plates, sixteen were initiated by photons, none by electrons. The theory also predicted that photons are generated in pairs by the decay of π°-mesons arising from nuclear interactions. A few of the cloud-chamber pictures appeared to verify this prediction by showing two electron showers originating from the same nuclear interaction and diverging at an angle consistent with the theoretical predictions.

While it is certainly true that charged and neutral π-mesons are the most common products of nuclear interactions, it is also true that, on rare occasions, nuclear interactions generate particles of a different kind, particles of which the 'new particles' discovered by Rochester and Butler, were the first examples. The search for these 'new particles', and attempts to establish their nature and behavior became one of the most fascinating and demanding activities of our group. As an example, I would like to mention an experiment performed by Annis, Bridge, Peyrou, and Safford at Echo Lake with a cloud chamber containing eleven lead plates, 0.6 centimeters thick. In a six-month operation, they obtained 12 000 pictures of particles stopping in one of the plates. Most of these particles could be identified as π- or μ-mesons. In six cases, however, this identification was not possible. The 'anomalous' particles were found to have a mass greater than that of π-mesons. After coming to rest they were seen to decay producing particles of mesonic mass. Therefore they belonged to the family of 'new' particles.

In the same set of photographs, sixty-two examples were found of a neutral particle decaying in flight, and producing two charged particles. Often one of the decay products could be identified as a proton, the other a meson. Thus it was possible to conclude that the neutral particle was a 'new' particle with a mass greater than the proton mass.

Fig. 5.8, finally, is the reproduction of an unusual event observed by our group, with a multiplate cloud chamber in 1954. It is of special interest because, in all probability, it is the first example of the annihilation of a negative proton with a positive proton (see figure caption for a detailed explanation). At a Conference of the *International Union of Pure and Applied Physics* (IUPAP), held in Bagnères-de-Bigorre in the summer of 1953 the results of observations of the 'new particles' obtained by our group and by other groups were of great interest. At the end of the meeting I was asked to summarize these results in a concluding speech. This was a difficult task because of the incredible variety of particles which the observations had revealed. There were certainly different kinds of 'new' particles heavier than protons, all of which, according to the nomenclature that came into use at the conference, became known in their ensemble as *hyperons*. There were also various kinds of 'new' particles of a mass intermediate between the meson and the proton mass; these were designated as *K-particles* (later they would become known as *heavy mesons*).

The conference of Bagnères-de-Bigorre was a landmark in the history of particle physics. The year of this conference was one of particular significance also for me personally. For in 1953 our third child, Linda, was born. Following an upstate New Yorker and a New Mexican, she was a living symbol of our wanderings.

Work on the 'new' particles continued for several years. It helped to set the stage of a new science to which, today thousands of workers and billions of dollars are committed.

Through the years, cosmic rays have been, to the physicist, an irreplaceable source of high-energy particles and to the cosmologist a fascinating, yet still undeciphered message from the heavens. The research program on nuclear interactions and elementary particles that I have just described was motivated by our interest in problems of physics. Interest in cosmological problems was the motivation of a second research program aimed, specifically, at the study of the primary cosmic radiation as a necessary step in any attempt of explaining the origin of cosmic rays.

One of the first experiments in this program was a search for electrons and/or photons in the primary radiation. The experiment was performed in 1948 by Robert Hulsizer, by means of a fast ionization chamber carried aloft by a string of neoprene weather balloons (large plastic balloons were not yet available). Flights were repeated several times. In some of them the chamber was unshielded, in others it was covered by a lead shield 2.5 centimeters thick.

From previous experiments we knew that the bursts observed in an unshielded chamber are produced almost entirely by the 'stars' arising from low-energy nuclear interactions of cosmic rays. The lead shield would not appreciably decrease the frequency of these bursts because 2.5 centimeters of lead is practically transparent to the star-producing radiation. However, if any high-energy electrons or photons are present, they will generate electronic showers in the shield which will produce additional bursts in the ionization chamber. Thus the difference between the burst frequency recorded with and without the shield will provide a measure of the number of electrons and photons at the location where the measurements are made.

Observations taken at an altitude of 30 kilometers failed to detect any appreciable number of electrons and photons; they were thus consistent with the view that the primary cosmic radiation does not contain any of these particles. A conservative estimate of the uncertainties in the interpretation of the experimental data made it possible to set an upper limit of one percent on the fraction of electrons or photons that might be present in the primary cosmic radiation.

Next, I wish to discuss in some detail a research program aimed at the study of extensive air showers, a program which, because of the originality of its conception and the significance of its results, ranks among the foremost accomplishments of the MIT group.

Extensive air showers are initiated by the arrival of primary cosmic-ray particles of exceptionally high energy, and develop in the atmosphere through a chain of nuclear interactions, electromagnetic interactions, and decay processes. As a result of these processes, the number of secondary particles increases, sometimes reaching values of billions or tens of billions. Scattering in the atmosphere spreads the particles over a large area, which becomes the effective area of a detector capable of recording the high-energy particles responsible for the production of the showers. The possibility of using extensive air showers as detectors of cosmic rays, in a range of energies inaccessible by any other means, was the main reason for our interest in this phenomenon.

As I have already noted, evidence for the existence of extensive showers had first been obtained in 1933, as a by-product of our experiments on the East–West effect performed in Eritrea. In 1938 Pierre Auger and his collaborators had started a research program which had confirmed the existence of extensive air showers and had also provided some information on their properties. This information, however, concerned the *average* properties of the showers, whereas we needed to know the *individual* properties of each

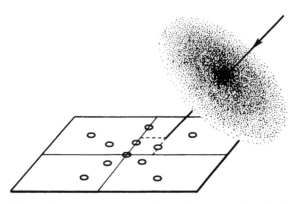

Fig. 5.9. Schematic representation of an extensive air shower approaching a detectro array. From *Cosmic Rays* (1964) by B. Rossi & G. Allen. Unwin Ltd, p. 187.

shower that was recorded. Specifically, we wanted to secure the information needed to determine the energy and direction of arrival of the particle which had initiated the shower.

Our plans for the measurement of the energy were based on a theoretical formula which relates the energy of the primary particle to the total number of shower particles at the location where the measurements were made. To determine this number, we had developed a method which we called the *density sampling method*. It involved the following steps: (1) measuring the density of the shower particles in a number of different locations; (2) finding, by means of these measurements, the center of the shower (i.e., the point of maximum density); and (3) constructing a curve giving the density of shower particles as a function of the distance from the center of the shower. With this curve we could immediately compute the total number of shower particles.

In order to determine the direction of arrival of the primary particle, which we assumed to coincide with the direction of the shower axis, we were planning to use a method, which we called the *fast timing method* based on a measure of the time intervals between the instants of time when the various detectors were struck by the shower front (see Fig. 5.9). The detectors needed to carry out the program outlined above had to satisfy two important requirements. Density sampling required proportional detectors, i.e., detectors whose signals are proportional to the ionization produced by the particles traversing their sensitive volume (and therefore to the number of these particles). Fast timing required detectors capable of measuring time intervals as small as one-thousandth of a second. It was also necessary that the detector's area be sufficiently large to allow measurements of small particle densities.

The G.M. counters that had been used in all previous experiments on extensive showers possessed none of these properties. In fact, at the beginning of our work in 1947, only one detector was known that would satisfy our requirements; this detector was the fast ionization chamber. Therefore fast ionization chambers were used by Robert Williams in the first experiment of our research program on extensive showers; measurements were taken with four chambers near Echo Lake and on the top of Mt Evans. The data, analysed by the density-sampling method, confirmed theoretical predictions concerning the dependence of the particle density on the distance from the center of the shower. They also proved that the primary cosmic radiation contains particles with energies in excess of 10^{16} electron volts.

Before turning to the problem of finding the directions of arrival of the primary particles responsible for the showers, it was necessary to verify whether the front of the shower was sufficiently sharp to permit the very accurate determination of its time of arrival on the detectors, as was required by the fast timing method. This verification was undertaken, in 1952, by George Clark, in collaboration with Piero Bassi from Italy who was then a guest of our group.

In the meantime, a new very useful research tool – the *scintillation counter* – had been made available to physicists. The development of this instrument was the product of recent studies which had shown that some fluorescent substances (both solid and liquid), when struck by ionizing particles, emit light flashes (scintillations) of only a few milliseconds duration. The operation of the scintillation detector is based on the use of one of these substances (scintillator) and of a photomultiplier which changes the light flashes into electric pulses. Like the fast ionization chamber, scintillation counters have a very fast and closely proportional response. But they are less sensitive to disturbances and much easier to operate. For these reasons, Clark and Bassi decided to use scintillation counters as detectors for their experiment.

The choice of the scintillating material presented a problem. Some fluorescent plastics had been found to be efficient scintillators. However sufficiently large samples of fluorescent plastics were not commercially available, and even small samples were prohibitively expensive. Thus using liquid scintillators was the only possibility. The scintillators that were employed in the preparation of the scintillation counters consisted of five-gallon drums filled with a solution of terpenyl in benzene (a mixture which had been found to have the properties of an efficient scintillator). Three scintillation counters were mounted on the roof of the Physics department building of MIT and arranged in various configurations during the different phases of the experiment. The observations occupied several months during the winter between 1952 and 1953. The results were gratifying, for they showed that, at any instant of time, practically all shower particles are contained in a disk of only one or two meters thickness which travels, with practically the speed of light, in the direction of the shower axis. This meant that the arrival time of the shower on a detector could be measured with the accuracy required by the fast-timing method.

Encouraged by these results, we started planning a major experiment no longer of an exploratory nature as the previous one, but aimed at providing detailed information on the extensive air showers. Through the study of air showers, we intended to investigate the energy spectrum of the primary particles from which they were produced, extending the observations to energies much greater than had been reached before. We were also planning to determine the directions of arrival of the primary shower-producing particles in order to discover any possible preferential direction that might offer a clue about the place of origin of the particles. Several members of the group became involved with me on this ambitious project; George Clark, William Kraushaar, John Linsley, James Earl, and Frank Scherb. In addition, Minoru Oda and Piero Bassi contributed to the preparation of the experiment before returning to their countries.

We knew that primary cosmic-ray particles of very high energy would be rare, and we knew that the large showers initiated by these particles would strike the ground sparsely over an extended area. Therefore, in order to observe a significant number of these showers in a reasonable time, it was necessary to distribute detectors of shower particles over a correspondingly extended area. To find a suitable free area in the vicinity of Cambridge was not easy. This problem was solved when, thanks to the kind intercession of the director of the Harvard Astronomical Observatory, Professor Donald Menzel, we were allowed to install our detectors on the grounds of the Agassiz Station. Part of the observatory, this station is located in a wooded region, some twenty miles from Cambridge.

In our experiment we started with fifteen scintillation counters with liquid scintillators of one square meter area. The counters were arranged in two concentric circles, with one counter at the center. The outer circle had a diameter of about half a kilometer. The signals of the fifteen photomultipliers were brought, with cables, to fifteen oscillographs and recorded photographically.

Observations began in July 1953 and, for a while, proceeded smoothly, except for occasional interruptions caused by the rabbits which liked to chew the cables. At the same time, we began studying methods for deriving the desired information about the properties of the showers from the observations. But shortly thereafter, when we had just recorded the first large showers, an incident occurred which threatened to bring the experiment to a halt.

At night, during a severe storm, lightning struck one of the scintillation counters, causing the liquid scintillator to burst into flame. Firemen arrived in a rush from a nearby town and promptly extinguished the fire. The trees soaked with water from the rain, did not catch fire, and the observatory building was spared.

Six months went by before, after many negotiations between MIT and Harvard, and after we had installed an elaborate system of fire protection, we were allowed to resume our experiment. But the incident had emphasized the advisability of replacing the flammable liquid scintillators with non-flammable solid scintillators. As I have already pointed out, samples of fluorescent plastics of sufficient size for our experiment were not available commercially. Therefore, if we insisted on using plastic scintillators, we had to learn how to manufacture them ourselves.

Clark took this task upon himself. In a few months, with the help of Frank Scherb and William Smith (our head technician to whom, for many years, we were indebted for valuable support of our work) he succeeded in starting a 'factory' capable of routinely producing disks of fluorescent plastics ten centimeters thick, and about one meter in diameter.

In the late spring of 1956 the liquid scintillators were replaced with plastic scintillators, and the observations were resumed. In a little over one year several thousand showers were recorded. The most important results of these observations included:

1 a precise measurement of the density of shower particles as a function of the distance from the shower center (a measurement that was in agreement with the theoretical predictions);
2 a measurement of the energy spectrum of the primary particles responsible for the showers from 10^{15} to 10^{18} electron volts;
3 the proof that these particles arrive in practically equal numbers from all directions;
4 the observation of a shower produced by a particle with an energy close to 10^{19} electron volts.

The isotropic distribution of the arrival directions of shower-producing particles suggested by the result of the Agassiz experiment was, clearly, a feature of the cosmic radiation of important cosmological significance. Therefore we felt that it should be tested by experiments, designed primarily at making directional measurements on large numbers of showers rather than at searching for the rare showers of the highest energy.

Thus, while the Agassiz experiment was still in preparation, Clark reactivated, with the addition of a fourth scintillation counter, the small array on the roof of MIT that he and Bassi had used previously. Observation of 2600 extensive showers, produced by particles with energies in excess of 10^{14} electron volts, did not provide any evidence for a preferred direction of arrival of these particles. Later, Clark, in collaboration with Vikram Sarabhai and Eknath Chitnis, performed a similar experiment at Kodaikanal, in India, at a latitude of 10° North. Their purpose was to find out whether, by any chance, the distribution of the arrival directions of the shower-producing particles near the equator was different from that observed in the temperate zone. In this experiment about 100 000 showers of a size corresponding to an average primary energy of 10^{14} electron volts were observed. Again, no evidence of an anisotropy was found.

Still another significant experiment in our research program was suggested by one of our foreign guests, Ismael Escobar. A political refugee from Franco Spain, he had settled in Bolivia where he had become a professor at the local University and the director of meteorological services. He had come to the United States on a Guggenheim Fellowship in order to acquaint himself with the research programs of his interest that were in progress at MIT and at other institutions.

Ismael pointed out that a location known as *El Alto*, at 4300 meters above sea level, on the Bolivian plateau near Le Paz was an ideal place for a high-altitude experiment on extensive showers. Electric power and other facilities were available. The level, unobstructed ground lent itself to the display of a large array of detectors. Climatic conditions allowed continuous observations throughout the year. On the basis of this information, we decided to move the instrumentation which had been used in the now completed Agassiz experiment to El Alto. A series of measurements taken by George Clark, Ismael Escobar, and Juan Hersil, one of Escobar's assistants, brought out interesting differences between the structure of the showers observed at El Alto and in Massachusetts; like previous experiments they failed to disclose any anisotropy in the directions of arrival of the shower-producing particles.

The El Alto experiment was the seed of the so-called *Bolivian Air Showers Joint Experiment* (BASJE), an international research program promoted by Escobar and based in a laboratory located near the top of Mount Chacaltaya, at an altitude of about 5000 meters. For several years the Japanese physicist Koichi Suga was effectively the soul of the BASJE program.

The success of the Agassiz experiment encouraged us to consider the possibility of another experiment with a detector array of much larger dimensions. Available astronomical information suggested that such an experiment might produce very significant results. Observational data indicate that our galaxy has roughly the shape of a flat disk, rather densely populated, surrounded by a diffuse halo. It was generally believed that cosmic rays came, at least primarily, from sources located in the disk, which means that the sources are not distributed symmetrically around the Earth. There was, however, no contradiction between the asymmetric distribution of the sources and the practically perfect isotropy of the observed cosmic radiation. It was known, in fact, that interstellar space is permeated by a magnetic field, which deflects the path of electrically-charged particles into complex curved trajectories, whose radius of curvature increases with the increasing energy of the particles. As long as this radius is small relative to the thickness of the galactic disk, the particles remain confined to the disk and strike the Earth without any memory of their place of origin. The isotropy of the observed radiation may, then, be explained as a consequence of confinement by the magnetic field. Confinement also has the effect of increasing the density of the particles in the disk and, consequently, of increasing the intensity of the radiation seen by a terrestrial observer.

When, however, with increasing energy, the radius of curvature approaches or exceeds the thickness of the galactic disk, then the particles will escape from it. In this case one may anticipate that very few of the particles from sources in the galactic disk will reach the Earth and that most of the particles observed at the Earth originate from sources outside of the disk. From the available information about the thickness of the disk (about 10^{21} centimeters) and about the average intensity of the interstellar magnetic field (about 5×10^{-5} gauss) one might predict the magnetic field should gradually lose its power to hold cosmic-ray particles within the galactic disk for particle energies within a range centered at about 10^{18} electron volts. Particular interest was attached to a study of cosmic rays in this energy range. Their flux was expected to be much weaker than the flux of the particles of lower energy recorded in the Agassiz experiment. Therefore their observation required a detector array of much larger dimensions.

For some time we debated whether it was advisable for the group to embark on the difficult project of building such an array. But eventually we felt that the potential results were worth the effort, and we decided to proceed. At his own request, the main responsibility for the project was assigned to John Linsley. The first problem was to find an area sufficiently large and free of obstacles to display this array. Obviously we would have liked to remain in the generally vicinity of MIT. But, after a careful search, we were forced to recognize that a site suitable for our experiment was not to be found in the northeastern part of the United States. We then decided to turn our attention to the wide open spaces of the American

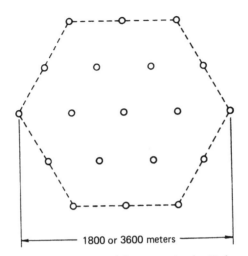

1800 or 3600 meters

Fig. 5.10. Hexagonal array of detectors in the Volcano Ranch experiment.

southwest. Eventually the choice fell upon a large semi-desert area, part of a property known as *Volcano Ranch* from three small extinct volcanos found within its boundaries. This property is located in New Mexico, about sixteen miles west of Albuquerque, at 1770 meters above sea level.

In the meantime, another physicist, Livio Scarsi, from the Occhialini school, had joined our group as a guest. Between 1957 and 1958 Linsley and Scarsi installed an array of 19 scintillation counters at Volcano Ranch. The array had the shape of a regular hexagon, subdivided into equilateral triangles, as shown in Figure 5.10. The hexagon was inscribed in a circle whose diameter was, initially, 1.8 kilometers, and was later increased to 3.6 kilometers. Each of the scintillators consisted of four disks of fluorescent plastics, similar to those used in the Agassiz experiment, which came from the 'factory' that George Clark had set up at MIT. The four disks were viewed by four photomultipliers connected in parallel.

The method used for recording and interpreting the experimental data was essentially the same as the one that had been used in the Agassiz experiment. Briefly, it may be described as follows. The signals of the nineteen scintillation counters were taken, through delay lines, to the vertical deflecting plates of nineteen separate oscillographs placed in a control cabin located in a wooden shack near the center of the array. The horizontal sweeps of the electron beams in the nineteen oscillographs were started simultaneously when at least three detectors were struck by an extensive shower. The delay lines, which I mentioned above, made sure that the signals from the photomultipliers reached the oscillographs after the start of the sweeps. The arrival of a shower struck the various detectors. The height of the peaks was a measure of the energy spent by the shower particles in the scintillators.

The first step of the analysis was a determination of the direction of the shower axis, by the fast-timing method. By measuring the height of the peaks, and taking into account the inclination of the shower axis with respect to the horizontal plane on which the scintillation counters were distributed, one could compute the density of particles at various points of the shower front. The density sampling method was then used to find the center of the shower, to construct the curve of the density of shower particles as a function of the distance from the center, and finally to compute the total number of particles in the shower. Regular measurements were initiated in September 1959 and continued until 1963. On frequent occasions during this period, I visited Volcano Ranch, motivated by a keen interest in the experiment and by a desire to maintain contact with my younger friends working there. A further inducement, I must admit, was my nostalgic feelings for New Mexico.

To keep the experiment working was not an easy task. The large dimensions of the array, and its isolated location, were, by themselves, serious obstacles. To reach the site of the experiment, starting from the state highway, one had to drive over eleven kilometers of desert, on an almost non-existent road, often in the midst of a snow- or sand storm. Additional trips in the desert were necessary to keep the operation of the detectors under surveillance, a precaution that was absolutely essential in order to ensure the reliability of the observational data. If one considers that there were almost 500 vacuum tubes, spread over a large desert area, one can easily appreciate the effort required by this work of surveillance. We also had to worry about spurious signals from a nearby radio station and from atmospheric electric discharges during the frequent storms in the summer; it had been necessary to install special devices as protections against these disturbances. Other causes of concern were the damage to the cables caused by ground squirrels and by the transit of cattle, and also the change in the electric resistance of the cables due to the violent thermal excursion in the desert: torrid temperatures in daytime, freezing temperatures at night. To protect the array from these troubles it was found necessary to bury the cables, another time-consuming job.

These difficulties, however, did not prevent the success of the experiment. From September 1959 to May 1960 Linsley and Scarsi carried out observations with the array in a circle of 1.8 kilometers in diameter. The largest shower recorded in this period contained thirty billion particles; the energy of the primary particles responsible for this shower was estimated to be sixty billion billions electron volts (6 × 10^{19} eV).

Later, after Scarsi had returned to Italy, Linsley single-handedly continued the experiment after increasing the diameter of the array from 1.8 to 3.6 kilometers. Intermittent observations lasted until 1969. The largest showers recorded in this second period contained fifty billion (5 × 10^{10}) particles; the corresponding energy of the particle responsible for its production was 10^{20} electron volts, or 6 joules, an almost incredible amount of energy for a particle of subatomic dimensions.

The comparatively large number of particles observed in the energy range up to 10^{19} electron volts made it possible to turn from the properties of the individual particles to the properties of the radiation as a whole. It was thus found that the energy spectrum of the radiation decreases rapidly but smoothly with increasing energy. It was also found that even in the very high energy range now being observed, no indication could be found for a preferential direction of arrival of the primary particles.

These were important, new results. But the most striking result was the proof that the primary cosmic radiation contains particles of an energy greater than the energy of the particles that could be contained in the galactic disk by the galactic magnetic field. Therefore, one had to infer that these particles came from sources external to the galactic disk; perhaps sources located in the galactic halo, or more likely located altogether outside of our galaxy.

6

Physics in space

At the end of the 1950s the appearance on the scene of new accelerators, capable of producing strong and controlled beams of high-energy particles, deprived cosmic rays of the monopoly over the study of high-energy nuclear interactions and of the elementary particles born of these interactions. Many other interesting aspects of cosmic-ray physics still remained to be explored. Among these were temporal intensity changes related to solar activity or due to other causes, high-energy extensive showers, etc. But these did not suffice to fill the vacuum left by the transfer of high-energy nuclear physics to accelerators.

As luck would have it, while the scope of cosmic-ray research was shrinking, the advance of space-flight technology was opening a rich, new field of scientific inquiry. Cosmic-ray physicists were in a privileged position to take advantage of this opportunity because of some kinship between cosmic-ray and space research with regard to both objectives and experimental tools. It is no wonder that these scientists should have been responsible for most of the early achievements of space science. My own contribution to this science was motivated by an interest in two subjects: the exploration of outer space and a search for celestial X-ray sources located outside the solar system.

Interplanetary space

Space activity in the United States, spurred by the success of Soviet space technology, was then in rapid expansion. In January 1958, barely four months after the launching of *Sputnik*, the first American satellite, *Explorer I*, went into orbit. With the detection of the Van Allen belt, this satellite started a long series of discoveries.

In the meantime the Federal government was proceeding to

completely reorganize the space program by assigning the control over all space activities of a non-military nature to a new agency, the *National Aeronautics and Space Administration* (NASA). For its part the National Academy of Sciences created a committee, the *Space Science Board*, whose assignment was to stimulate and co-ordinate scientific activity in space. Initially the Board, under the chairmanship of Lloyd Berkner, included sixteen scientists, chosen from among the experts in one or another of the disciplines which were expected to substantially advance from observations in outer space. I was asked to join the Board, even though I was, in a way, a foreigner in the group. Or perhaps this was just the reason why I had been invited, my task being to discover possible gaps in the program developed by the experts.

Each Board member was asked to form a subcommittee. Three eminent scientists and close friends of mine – the biologist Salvador Luria, the physicist Philip Morrison, and the astrophysicist Thomas Gold – agreed to join my subcommittee. The subcommittee held its first meeting in September 1958. One of the conclusions reached at this meeting was the desirability of initiating a program aimed at the exploration of the physical conditions of interplanetary space.

As a matter of fact, it was hard to understand why this explora-tion should not have been included in the early program of NASA. Actually for several years observations from Earth had suggested that the space surrounding our planet is not entirely devoid of matter, as had been supposed in the past but contains a dilute *plasma*, i.e., an ionized gas consisting, presumably, of electrons and protons. Thus, already in 1930, Sidney Chapman and Vincenzo Ferraro in England had advanced the hypothesis that magnetic storms are produced by streams of ionized particles coming from the Sun and directed toward the Earth. More recently, in 1950, Ludwig Bierman in Germany had shown that the tails of type 1 comets (those formed by electrons and ionized molecules) could not be produced by the pressure of solar light (as had been assumed until then), and had suggested that the agent responsible for the formation of the tails was a fast stream of ionized gas originating from the Sun. In addition to these (and other more questionable) pieces of evidence derived from observations, theoretical considera-tions also supported the view that the space around the Earth contains a plasma in motion. According to a theory developed by

Eugene Parker, a physicist at the University of Chicago, the solar corona is not in a state of stationary equilibrium but expands steadily outward thus producing a plasma wind in interplanetary space, which Parker called the *solar wind*.

However, while there was general agreement about the presence of a plasma in interplanetary space, the views concerning the properties of this plasma were widely divergent. The estimates of the plasma density ranged from one to one thousand electrons and protons per cubic centimeter. Some believed that the plasma was nearly stationary, others that it was flowing with a speed of 1000 kilometers per second. Perhaps the plasma was distributed more or less evenly in space; perhaps it was condensed into clouds.

For us in the cosmic-ray group the problem of a plasma in interplanetary space was nothing new. For some time, we had been concerned with this problem because of the possibility suggested by some scientists that certain temporal changes of cosmic-ray intensity might be due to clouds of magnetized plasma ejected by the Sun into the surrounding space. I well remember the help we received from Philip Morrison and Thomas Gold in our attempts to bring the problem into focus. Therefore, having dutifully reported to the Space Science Board the recommendations of my subcommittee, I felt motivated to initiate a study of interplanetary plasma with my own group at MIT. Thus was born a research program which continues to this day and has been one of the major research activities of the Institute.

The first to join this new venture was Herbert Bridge who would later become one of the leaders in the science of interplanetary plasma. Others followed: Frank Scherb, Edwin Lyon, Alan Lazarus, Constance Dilworth-Occhialini, Alberto Bonetti, and Alberto Egidi.

Our plan called for an experiment to be performed in outer space by means of a plasma probe carried aloft by a satellite. In designing the probe, several requirements had to be kept in mind. In the first place, since one of our purposes was to study the motion of the plasma, the probe must record the protons and ignore the electrons, because in a fast-moving plasma made of protons and electrons, the protons form a beam of particles all moving in almost the same direction, which is the direction of motion of the plasma itself, whereas the electrons, on account of their high thermal agitation,

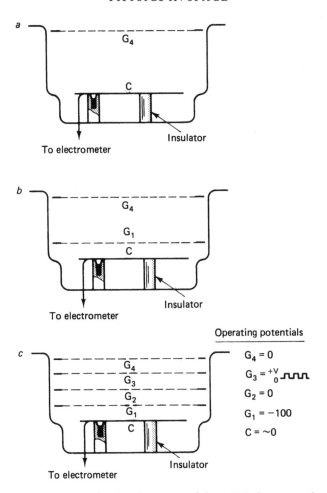

To electrometer

Insulator

To electrometer

Insulator

Operating potentials

$G_4 = 0$

$G_3 = {}^{+V}_{0}$ ⊓⊔⊓⊔

$G_2 = 0$

$G_1 = -100$

$C = \sim 0$

To electrometer

Insulator

Fig. 6.1. Steps in the development of the MIT plasma probe.

move more or less at random and therefore cannot provide any information on the bulk motion of the plasma. Second, because of the very large uncertainty concerning the properties of the plasma, a large dynamic range was an essential property of the measuring instrument. Finally, we wanted our probe to be of solid design, simple in its construction and fully reliable in its operations. With these requirements in mind, we planned our probe starting from the model of a classical, well-tested instrument, the *Faraday cup*. We would build this cup (see Fig. 6.1*a*) by placing a plate C (the collector), behind a grid, G_2, that covers a hole in the skin of the satellite. The grid would be kept at the potential of the satellite's skin (to be called *zero potential*) while the collector would be brought to a negative potential. When a plasma stream entered the chamber through the grid, the protons would reach the collector, while the electrons would be repelled by it. Thus in the collector circuit there would appear an electric current of an intensity proportional to the proton flux.

However, we were aware that measurements taken with an instrument of this design would be subject to a serious source of error, because the strong solar radiation impinging upon the collector would eject electrons which, moving toward the grid, would produce in the collector circuit an electric current of the same sign as that produced by the protons moving in the opposite direction. We could remove this inconvenience by placing, in front of the collector, a second grid, G_1, kept at a negative potential relative to the collector (see Fig. 6.1b). This grid, by repelling the photoelectrons emitted by this electrode toward the collector, would stop the *direct* photoelectric current created by these electrons. While doing so, however, it would generate a *reverse* photoelectric current due to the solar radiation reflected toward it by the collector. We soon realized that it was not possible to entirely suppress the photoelectric disturbances by adding more grids, because each new grid would have been a new source of photoelectrons. Clearly, we had to contrive some entirely different method of making the response of our probe insensitive to the photoelectric effect.

It was not an easy problem. But finally we found a satisfactory solution, which consisted of *modulating* the proton flux *selectively*, i.e., without, at the same time, *modulating* the photoelectric current. The modulation was achieved by a third grid (the *modulating grid* G_3 in Fig. 68.1c), whose electric potential oscillates rapidly between zero and some positive voltage (the *modulating voltage*). The modulating grid will cut off the flow of protons with energies below a certain value which, for normal incidences, equals the value of the modulating voltage (if this voltage is measured in volts, and the energy is electron volts). The modulation of the proton flow will produce in the detector circuit an alternating current of an intensity proportional to the modulated part of the proton flux. Photoelectric currents, not subject to modulation, will not affect the measurements.

We were concerned about the possibility of disturbances arising from a capacity coupling between the modulating grid and the collector. Actually, on the face of it, we did not see how it would be possible to record millivolt signals in the vicinity of an electrode where voltage fluctuations of hundreds or thousands of volts were taking place. Thus I did some figuring and found that while, in fact, a single grid between the modulating grid and the collector (such as the grid G_1 already included in the design of the probe) would not suffice to entirely shield the collector, two grids would afford perfectly adequate protection. On the basis of this result, we decided to insert still another grid (G_2) directly behind the modulating grid.

Having completed the conceptual design of the probe, we proceeded to construct a model of the instrument, which we then submitted to a number of tests. For this purpose, we placed the probe in a vacuum chamber and exposed it to beams of protons of different energies, incident at different angles with respect to the normal to the cup. We used a modulating frequency of 1400 cycles per second. For each energy of the protons and for each angle of incidence, we measured the alternating current in the collector circuits, with values of the modulating voltage ranging from 5 to 2300 volts. The results were in perfect agreement with the expectations. The minimum observable current turned out to be of about 2×10^{-11} amperes, corresponding (for protons at normal incidence) to a flux density of $4 \times 16^6/cm^2/s$. The maximum detectable flux density (a limit set by the saturation of an electronic amplifier in the collector circuit) was about 2×10^{10} protons cm^2/s.

Our tests confirmed the expectation that the probe would have the capability of measuring the energy spectrum of the plasma protons, a property that derived from the fact that the intensity of the alternating current in the detector circuit was a measure of the number of protons in that part of the energy spectrum which extends up to the energy of the protons subject to modulation by the applied modulating voltage. We also verified that no spurious signals occurred because of a residual coupling between the modulating grid and the collector. Most importantly, we verified that photoelectric currents produced by intense ultraviolet radiation did not in the least affect the alternating current in the collector circuit. These tests made us confident that we entirely understood the behavior of our probe in all of its details, and that there would not have been any ambiguity in the interpretation of its results. Having reached this stage of our work we were ready and anxious to fly.

While our work was in progress, other scientists were engaged in programs directed, like ours, to the exploration of physical conditions in interplanetary space. The only team which had already done some direct observations in space was a group of Soviet scientists, under the leadership of Konstantin Gringauz. From 1959 to 1961 they flew their detectors on four satellites, the first three directed at the Moon, the fourth at Venus. As implied in the name *charged particle traps*, these detectors were not designed specifically as plasma probes, but were to be used to measure fluxes of charged particles (electrons or protons) in outer space, and actually many of the results of the Soviet scientists referred to the electron population in the general proximity of the Earth. Like our probes, the Soviet traps were modified Faraday cups. But, as the Russian scientists were aware, their design did not provide adequate protection against photoelectric interference, nor did it afford the possibility of significant measurements of the proton energies, the two important features that modulation had provided for our probe. The Soviet observations of greatest interest to us were obtained by the traps aboard *Lunik III* during its flight to the Moon. From a geocentric distance of about 255 000 kilometers and up to impact on the lunar surface, the traps recorded a flux of charged particles, presumably protons. The authors correctly claimed that their experiment had provided the first evidence for the presence of a stream of charged particles in interplanetary space. However they could only state that the energy of the particles was greater than 15 electron volts, and that the direction of their motion was consistent with the assumption that they came from the Sun.

In the USA two other groups besides ours were making preparations for experiments on interplanetary plasma. The first was a group working at the Jet Propulsion Laboratory (JPL) under the direction of Conway Snyder and Marcia Neugebauer; the second was a group working at the Ames Laboratory of NASA, under the direction of Michael Bader. In their final design, the instruments developed by these two groups were based on the deflection of a stream of charged particles in the electric field between two curved plates held at different electric potentials. Despite the friendly personal relations between the members of the three groups, it was inevitable that when the time came to fly our instruments, a keen competition would develop for the privilege of being the first to do so. We felt that we deserved this privilege. I had been the first to alert NASA, through the Space Science Board, about the importance of plasma measurements in interplanetary space; with a series of lectures at various institutions I had done my best to bring the problem to the attention of the scientific community in the United States. Moreover our group had already developed and fully tested an original, most reliable plasma probe, capable of precise measurements.

Eventually, after several ups and downs, it looked as if our way was clear to an early flight. It so happened that, for some reason, NASA had changed the mission of one of its satellites, *Explorer X*, from a flight to the Moon to an orbit around the Earth. James Heppner, a physicist working at the Goddard Space Flight Center of NASA, obtained the use of *Explorer X* for a study of the geomagnetic field in the space around the Earth. Clearly, simultaneous measurements of plasma and of magnetic field were essential for a complete description of the hydromagnetic conditions in outer space. Thus our request to fly our plasma probe on the same satellite was accepted. But our worries were not yet over. For we learned that a plasma probe, developed by another laboratory, might fly before ours on a different satellite. At this point (we are now in the spring of 1960) I wrote a long letter to Homer Newell, Deputy Director of Space Science at NASA which read, in part: 'The interest in a program of plasma measurements that we have succeeded in generating has not been sufficient to ensure adequate support to this program, but might well have been sufficient to freeze us out of it.' These were harsh words, explained, if not justified, by my strong concern over the fate of our experiment. I seem to have felt that I was not being fair to NASA and, in particular, to Homer Newell (our best friend at NASA Head-quarters) for I ended the letter with the words: 'I realize that I have disposed of in less than three lines of our debt of gratitude to NASA, while I have filled more than two pages of criticism. This, I am afraid, is the incurable ingratitude of human nature.' I do not know whether this letter had any effect. Be this as it may, on March 21, 1961, *Explorer X*, carrying our plasma probe and Heppner's magnetometers, was launched into a very elongated elliptical orbit, which reached an apogee of 46.6 earth radii, in a direction about 33° from the anti-solar direction.

The magnetic field detectors included a rubidium vapor magnetometer and two flux gate magnetometers. Our plasma probe was mounted on the side of the satellite, looking in a direction perpendicular to the satellite's axis (see Fig. 6.2). An aspect sensor, provided by the Goddard Space Flight Center, was used to indicate the instantaneous orientation of the satellite.

During the flight, the spacecraft was spin-stabilized and spun around its axis with a rotation period of 548 milliseconds. It was anticipated that, if the plasma had a sufficiently high bulk velocity relative to the satellite, the proton flux detected by the probe would be modulated by the satellite's rotation, reaching a maximum when the direction of the normal to the cup came closest to the plasma velocity vector. The sharpness of the maximum would provide a measure of the degree of collimation of the proton beam. Solar cells were not yet available. Thus our satellite was powered by chemical batteries which provided reliable operation of the probe for about sixty hours, during which time the satellite almost reached the apogee.

The International Conference on Cosmic Rays and the Earth Storm, held in Kyoto in September 1961, provided the opportunity of acquainting the scientific community with our *Explorer X* experiment and of presenting some preliminary results. The most significant finding was the existence of two sharply separate regions around the Earth. In the first region (nearer to the Earth) the plasma probe did not give any signal that could be ascribed to an interplanetary plasma (while the magnetometers recorded a fairly regular magnetic field). In the second region (farther from the Earth) the plasma probe recorded a substantial, although variable, proton flux, while the magnetometers recorded a somewhat weaker, irregular magnetic field. The proton flux was strongly modulated by the rotation of the satellite, the maximum intensity being observed when the angle formed by the normal to the cup with the Sun's

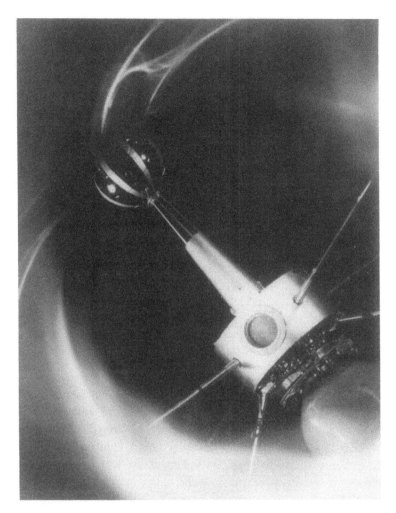

Fig. 6.2. The MIT plasma probe mounted on the side of *Explorer X*.

direction went through a minimum (see Fig. 6.3). The average energy spectrum of the protons peaked around 500 electron volts. The typical density was in the range of 6 to 20 particles per cubic centimeter. *Explorer X*, travelling in a direction away from the Sun, crossed the boundary between the two regions at a distance on the order of twenty Earth's radii.

In the months following the Kyoto meeting, we carried out a

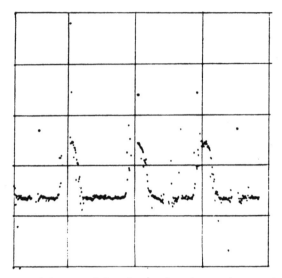

Fig. 6.3. Modulation of the plasma flux due to the rotation of the satellite. Vertical lines show the azimuth of the Sun. From an article by A. Bonetti, H. Bridge, A. Lazarus, E. Lyons, G. Rossi & F. Scherb in *Space Research*, III, 543 (1963).

detailed and critical analysis of our plasma measurements. We compared them with the simultaneous measurements of the magnetic field and examined the consequences that could be drawn from the experimental data. At the Third International Space Science Symposium held in Washington in the spring of 1962, members of our group described the experimental procedure and the final results of our work. In a speech summarizing our conclusions I said:

Behind the Earth (i.e., downstream with respect to the plasma wind) there exists a region which is effectively shielded from the wind by the Earth's magnetic field. The boundaries of this region, which we may call '*geomagnetic cavity*', appear to be quite sharp. Beyond them, a plasma flow is observed, whose protons have a mean kinetic energy of about 400 ev, indicating for the plasma a bulk velocity of about 300 km per sec. The proton flux fluctuates around a mean value on the order 3×10^8 particles per cm^2 per sec, indicating a plasma density on the

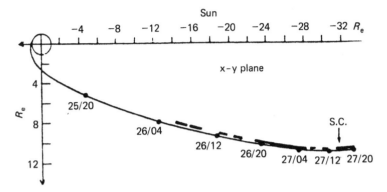

Fig. 6.4. Projection on the plane of the ecliptic of the trajectory of *Explorer* X. Heavy lines show the sections of the trajectory where plasma was observed. From a paper by A. Bonetti, H. Bridge, A. Lazarus, E. Lyons, B. Rossi & F. Scherb in *Space Research*, III, 542 (1963).

order of 10 protons and electrons per cm^3. The direction of the plasma wind lies within a 'window' of about 20° × 60° aperture, which includes the Sun. An appreciable energy spread is observed, which may be explained by the assumption that the moving plasma has a 'temperature' between 10^5 and 10^6 degrees Kelvin. Explorer X crossed the boundary of the geomagnetic cavity at a distance of about 22 Earth radii from the center of the Earth. However, on several occasions during the rest of the flight, the plasma current disappeared and then reappeared again. A tentative interpretation of this effect is that the satellite was flying close to the boundary of the geomagnetic cavity and that this boundary was not fixed in space but was moving back and forth, perhaps as a consequence of variations in the speed or in the density of the plasma wind.

Figure 6.4 shows a projection of the *Explorer* X trajectory on the ecliptic plane. The heavy segments indicate the sections of the trajectory where substantial fluxes of protons were detected.

Comparing the velocity of the plasma flow with the velocities of plasma waves, we found that the plasma wind was supersonic. Therefore we concluded that:

The picture emerging from *Explorer* X data is that of a

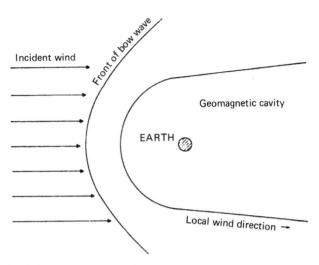

Incident wind

Front of bow wave

Geomagnetic cavity

EARTH

Local wind direction →

Fig. 6.5. Schematic picture of the geomagnetic cavity and the bow wave. From an article by A. Bonetti, H. Bridge, A. Lazarus, B. Rossi & F. Scherb in *Journal of Geophysical Research*, **68**, 401 (1963).

supersonic flow of plasma around an obstacle represented by the Earth's magnetic field. Under these circumstances one should expect the formation of a 'bow wave', pointing towards the direction of the oncoming wind and enveloping the geomagnetic cavity.

A schematic illustration of the above conclusions appears in Figure 6.5.

We were aware of the possibility that *Explorer X*, while outside the magnetic cavity, might have remained inside the bow wave, and that the properties of the plasma might have been modified by the transversal of this surface. However, we took the view that, whatever interaction with the Earth's magnetic field might have occurred, it could not have increased the wind's velocity; thus from the supersonic character of the observed plasma flow, it was safe to conclude that the plasma flow in free space was also supersonic. (Actually, the observed velocity could not have been very different from that of the wind in free space because *Explorer X* was flying along the tail of the geomagnetic cavity, which was nearly parallel to the plasma flow.)

In the years that followed the flight of *Explorer X*, the study of the interplanetary plasma and of its interactions with the various bodies of the solar system has become one of the most active branches of space research. Observations were carried out at greater and greater distances from the Earth, with flights to the Moon, then to Mars, Venus, Mercury, Saturn, Uranus, and Neptune. The MIT group, under the leadership of Herbert Bridge, and with the active participation of other experimental and theoretical physicists, foremost among them Stan Olbert, has kept at the forefront of these research activities. The modulated plasma probe, originally developed for the *Explorer X* experiment is still being used, having proven to be the most effective detector for many kinds of plasma measurements.

X-ray astronomy

In 1959, when I first began to think about the experimental program which was to become the forerunner of X-ray astronomy, the Sun was the only known celestial source of X-rays. It is a comparatively weak source, which can be detected easily from the Earth because of its proximity. For over ten years, solar X-rays had been studied extensively by scientists at the Naval Research Laboratory (NRL), under the leadership of Herbert Friedman. Astronomers had known that it would be very desirable to extend their observations of more distant celestial objects into the X-ray range of the spectrum, but they had been deterred from doing so by the foreseeable extreme weakness of the X-ray signals reaching the Earth from these objects.

Because of the strong absorption of X-rays in air, extraterrestrial X-rays can be detected only from above the atmospheric blanket. For their observations of solar X-rays the NRL team had used detectors carried aloft by the rockets that were then available. Now the more powerful rockets and the satellites, products of the developing space technology, had made it possible to hold the detecting instruments outside the atmosphere for much longer periods of time than could have been done before, a prime requirement for the detection of much weaker X-ray sources.

I felt induced to exploit this capability of space vehicles in an attempt at observing X-ray sources located outside the solar system.

I discussed this idea with my colleagues of the MIT cosmic-ray group. They were greatly interested, but all of them were heavily committed to other important research projects: the study of the interplanetary plasma (which I have already described), and a search for gamma-rays from celestial sources, which had been initiated by William Kraushaar and was now pursued by him in collaboration with George Clark.

I had to agree that, under these circumstances, the group was not in a position to undertake another major research program. But still I was not ready to forgo my plan. Thus in early 1959 I went to visit Martin Annis, an old friend and former student of mine, who in 1958, with support of some friends, had established a company, American Science and Engineering (AS&E), specialized in research and development. The strength of AS&E lay primarily in the quality of its scientific and technical staff. First Jack Carpenter had joined the company in the capacity of chief scientist. He was soon followed by several other scientists; among them Riccardo Giacconi, Frank Paolini, and Herbert Gursky. In my spare time from MIT duties, I served as Chairman of the Board of Directors and as scientific consultant to AS&E. George Clark, Stan Olbert, and Herbert Bridge were also acting as consultants.

During my conversation with Martin, I brought forward the possibility that his company might sponsor the project I had in mind. I did not try to minimize the difficulties of detecting celestial X-ray sources located outside the solar system. But I also stressed my belief that, if these difficulties could be overcome, the experiment would produce a result of historic importance, for it would open a new window on the Universe. Martin reacted enthusiastically to my proposal. He fully realized (as I did) that his fledgling company would take a considerable financial risk by committing a sizable part of its facilities to a venture of major proportion, without any apparent commercial applications. Nonetheless, he accepted the challenge.

To start, AS&E agreed to sponsor a study period intended, on the one hand, to critically examine the astrophysical information pertinent to our program, and on the other hand to plan the experimental technologies that would have to be developed for carrying out this program. This study was to be performed jointly by the AS&E scientists and by consultants from the MIT cosmic-ray group, including, besides myself, George Clark and Stan Olbert.

Clearly, the basic obstacle was the supposed small intensity of the X-ray signals originating from remote X-ray sources. The Sun, if removed to a distance comparable to that of the nearest stars (about four light-years) would send a flux of X-rays to the Earth over one hundred billion times weaker than it does from its actual position (about eight light-minutes). The Sun, of course, is just an ordinary star. Other less common celestial objects (supernova remnants, flare stars, etc.) might be much stronger X-ray sources. In particular, it was believed that the Crab Nebula, a supernova remnant, should be an especially strong X-ray emitter. Stan Olbert accurately analysed the physical processes likely to occur in the Crab and estimated the X-ray flux which they would generate. He came to the conclusion that any X-ray emission by the Crab was probably far too weak to be detected from Earth using existing instruments. A similar conclusion also applied to other potential extrasolar X-ray emitters. It thus appeared that a necessary requirement of X-ray astronomy was the development of X-ray detectors much more sensitive than the ones available at that time.

The detectors used by NRL scientists were G.M. counters provided with thin mica windows of about one square centimeter in area which allowed X-rays to enter their sensitive volume. The most obvious method to increase the sensitivity of a counter was to increase the size of its X-ray transparent window. But the sensitivity that could be achieved in this manner was limited by the cosmic-ray background. X-ray detectors are also sensitive to cosmic rays, and an X-ray signal can be perceived only if it is not swamped by the cosmic-ray signal. The response to X-rays depends on the area of the window. Since it is not possible to increase this area substantially without increasing the dimensions of the counter, thereby increasing its sensitivity to cosmic rays, it is clear that no substantial increase in the effective X-ray sensitivity of a counter can be achieved by an increase of its window area.

By considering various ways of overcoming the limitation to the sensitivity imposed by cosmic rays, we came to the conclusion that the most effective method was to selectively concentrate over a very small X-ray detector the X-rays incident over a large area of collection. The problem of concentrating an X-ray beam was not a new one. Several methods for achieving this purpose had been suggested and tested experimentally. One of them was based on the diffraction of the X-ray beam passing through a so-called *zone-plate* (a series of thin annular bands alternate opaque and transparent). We examined the zone-plate method in some detail, and decided that it was not suitable for our purpose.

Eventually, Riccardo Giacconi came forth with a promising proposal. He suggested to base the construction of an X-ray concentrator on the well-known property of X-rays to undergo total reflection when incident over a smooth metal surface at a grazing angle, i.e., at a small angle with respect to the surface itself. He pointed out that a parabolic mirror, used to reflect X-rays at grazing incidence, would perform as an effective concentrator. Shortly thereafter, Giacconi found an article by H. Wolter written in 1952 demonstrating theoretically that true X-ray images could be formed by a grazing incidence reflection over a parabolic mirror, followed by a second grazing incidence reflection over a hyperbolic mirror. Wolter had been aiming at the development of an X-ray microscope. Giacconi realized the possibilities that Wolter's scheme opened for the construction of an X-ray telescope, and became interested in taking advantage of these possibilities.

Wolter's theoretical results on grazing-incidence optics and on their applications to a possible X-ray telescope were discussed at some length by Giacconi and other scientists at a conference on X-ray astronomy held at the Smithsonian Astrophysical Observatory in May 1960, under the chairmanship of Albert Baez, a distinguished astrophysicist (and, incidentally, the father of Joan). It was clear that an X-ray telescope would be a valuable astronomical instrument. It was equally clear, however, that its development would require a lengthy effort. Therefore, it was held desirable to make a preliminary X-ray survey of the sky with some instrument easier to build, although necessarily less sensitive than a telescope.

I, personally, was greatly interested in this project. I was confident that, without resorting to any fundamentally new technology, it was possible to develop X-ray detectors substantially more sensitive than those previously used for solar observations. To be sure, the sensitivity of these detectors would still be a far cry from that deemed necessary for the detection of remote X-ray sources.

But no one had yet explored the sky with X-ray detectors as sensitive as those that I hoped could be developed, and this, for me, was a sufficient reason for undertaking this exploration; my long experience as a cosmic-ray physicist had taught me that when one enters an unexplored territory there is always a chance that he may find something unpredicted.

Besides this, if you wish, dream-like reasoning of mine, there was a more factual motivation for the experiment. At that time there were good reasons to suppose that the Moon should be a source of X-rays. Two different processes were thought to be capable of producing X-rays on the lunar surface. The first was a scattering of solar X-rays. The second was the impact of high energy electrons originating from the Sun (a process similar to that responsible for the operation of ordinary X-ray tubes). Stan Olbert had analysed these phenomena and reached the conclusion that the Moon was likely to be an X-ray source of sufficient strength to be observable from the Earth with detectors of modest sensitivity. Thus, the Moon was a convenient target for a first experiment on X-ray astronomy.

The possibility of observing X-rays from the Moon was, for us, a matter of only moderate interest. But this possibility had an important practical consequence. It so happened that the Cambridge Research Laboratories of the US Air Force (AFCRL) were engaged in a program of lunar studies. Therefore, they looked favorably on a project aimed at detecting an X-ray emission by the Moon and agreed to finance our experiment, in the hope that it would provide some information on the chemical composition of the lunar surface. I wish to point out that, while our experiment was to be officially aimed at the Moon, we hoped that it would also represent a first tentative step in the search for other much more remote X-ray sources. Thus, instead of using a detector with a restricted field of view, which, mounted on a spinning rocket would repeatedly scan a narrow strip of the sky containing the Moon, we decided to leave the field of view of the detector completely open so as to enable it to scan a large section of the sky.

The first task was to design and construct the detectors to be used in our experiments. Frank Paolini took this on. The detectors he produced, after a fairly considerable amount of work, were flat G.M. counters, each provided with seven mica windows having a total area of 20 square centimeters. An anti-coincidence system was

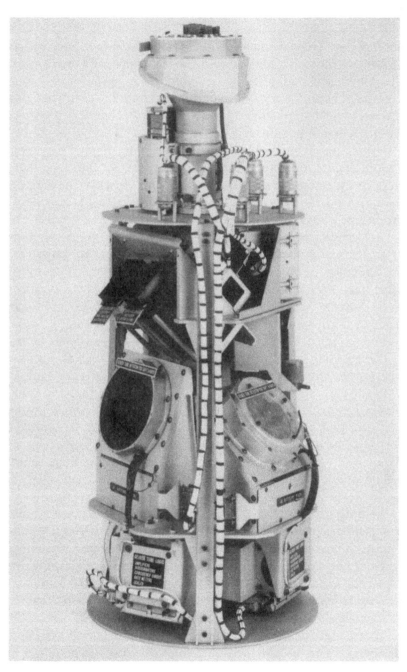

Fig. 6.6. Photograph of the rocket which discovered the first extra-solar source of X-rays. The rocket carried three flat G.M. counters, two with mica windows 0.002-inch thick, one with a mica window 0.001-inch thick.

used to reduce the cosmic-ray background. With this protection, the sensitivity of the counters was about one hundred times that of any X-ray detector used previously in rocket experiments.

A trial experiment was performed in November 1961 by means of a rocket equipped with 'conventional' X-ray counters. No results were obtained because of a mechanical failure of the rocket. For a second flight, scheduled for the following spring, the Cambridge Research Laboratories provided a more reliable and powerful rocket. Three of Paolini's counters were mounted symmetrically on the side of the rocket, facing in a direction at 55° from the rocket's axis (see Fig. 6.6). The rocket also carried an optical aspect sensor and a magnetometer.

The rocket took off on June 18, 1962, a day after full moon, from the White Sands missile range in southern New Mexico. It rose to a maximum altitude of 225 kilometers, while rotating at the rate of 2 revolutions per second around its spin axis. At each revolution the counters scanned a portion of the sky 100° wide. Throughout the flight the axis was pointing approximately in the vertical direction. At an altitude of 80 kilometers, the rocket having reached beyond the dense layers of the atmosphere, an increase in the pulse frequency of the counters signalled the appearance of a rather strong flux of radiation, presumably of X-rays. The signal was strongly modulated by the rotation of the rocket, evidence that the X-rays did not come in equal numbers from all directions. The rocket remained above 80 kilometers for about six minutes. Of the three counters, two worked reliably. Most significant were the data obtained by the counter with the thinner window (0.2 mil.). The dots in Figure 6.7 show the total number of counts recorded by this counter as a function of the azimuthal angle of the direction perpendicular to its plane. Thus, the curve outlined by these points represents the azimuthal dependence of the pulse rate. Similar, although less precise, were the data obtained by the other counter provided with a thicker window and therefore less sensitive to X-rays.

A careful analysis, taking into account the response of the counter to X-rays incident in different directions, showed that the shape of the curve was exactly that predicted for the case that the observed X-rays originate from a localized source, whose azimuth coincided with the azimuth of the curve maximum. Obviously we

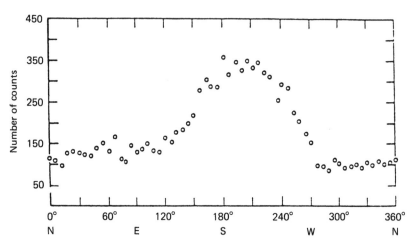

Fig. 6.7. Numbers of counts recorded by one of the counters with a thinner window as a function of the azimuth of the line of sight of the counter. From an article by R. Giacconi, A. Gursky, F. Paolini & B. Rossi in *Physical Review Letters*, **9**, 439 (1962).

first thought that the source was the Moon. But then, an accurate correlation between the data of the counter and those of the optical aspect sensor, proved conclusively that the azimuth of the maximum was at a distance of 30° from the azimuth of the Moon. Therefore the observed X-rays did not come from the Moon. Similarly, we could prove that they did not come from any other celestial object belonging to the solar system. We were thus led to presume that *the X-ray source was located outside the solar system.*

We noted that the curve in Figure 6.7 did not go exactly to zero on either side of the maximum. We thought that this 'tail' might be due to a diffuse celestial source of X-rays (whose existence, in fact was established by later observations). We ourselves did not yet fully believe the significance of our results which seemed to substantiate our wildest dreams. We spent weeks and weeks in an effort to make sure that nothing might have misled us in the interpretation of our observations. We meticulously analysed the response of our counters to agents that might have produced unwanted signals (cosmic rays, electrons spiralling around the lines of force of the Earth's magnetic-field; optical or ultraviolet rays from some auroral activity, etc.). Eventually we were able to satisfy ourselves that none of these phenomena could have simulated the X-ray beam which appeared to have been recorded by our detectors.

In the autumn, being finally certain that we had correctly interpreted our observations, we felt entitled to inform the scientific community of our discovery. We submitted a letter to the Editor of *Physical Review*, signed by Riccardo Giacconi, Herbert Gursky, Frank Paolini, and myself, ending with the words '*we believe that the data can best be explained by identifying the bulk of the radiation as soft X-rays from sources outside the solar system*'. As a further verification of our conclusion we repeated the experiment, in exactly the same conditions, in October 1962, when the presumed source was below the horizon, and again in June 1963. As expected, in October the maximum in the azimuthal distribution of the counting rate had disappeared; in June 1963 the maximum had reappeared in exactly the same position where it had been observed one year earlier.

In the meantime, in April 1963, Herbert Friedman's group at NRL had confirmed the existence of the X-ray source which we had discovered, and had located it in the constellation of Scorpio. Hence the name of Sco X-1, by which this source became known. (For the record, I may mention that, in a prior, private conversation, Frank Paolini had mentioned that from a careful analysis of the data of the June 1982 flight, he had concluded that the source was in the constellation of Scorpio. I regret that I did not encourage him to publish this result.)

The announcement of the discovery of an X-ray source located outside the solar system, presumably at a distance comparable to that of the stars was received with much interest but, at first, with considerable skepticism; so much so, that our 'letter' was at first rejected by *Physical Review*, and was finally accepted for publication only after I told Samuel Goldsmith, at that time the editor, that I assumed personal responsibility for its contents. The skepticism was justified. The existence of a celestial, presumably very remote X-ray source of such an extraordinary intensity as to be observable with our simple instrument appeared to be incompatible with all astrophysical notions. Actually the incompatibility was only apparent. In fact, the astronomical information available at that time concerned the celestial bodies which were then known. What our experiment had done had not contradicted this information, but had demonstrated the existence of previously unknown celestial objects to which such information did not apply.

As the last doubts about the credibility of our discovery were dispelled, and the awareness grew that observations in the spectral range of X-rays could discover vital features of our universe that were not accessible to optical or radio observations, increasing numbers of scientists were drawn into the field of X-ray astronomy. At MIT, first George Clark and Minoru Oda, then Hale Bradt, Gordon Garmire, and Saul Rappaport, turned from cosmic rays to X-ray astronomy. Through the years, the group that was thus formed attracted other scientists. Among them, Walter Lewin, Claude Canizares, and George Ricker, who are still present and are now senior members of our group. Others have left to form new groups at other institutions. The imaginative and strictly reliable character of its scientific contributions has earned for the MIT group an outstanding position in the scientific community. By the quality of its work, by the closely-knit fabric of the relations between its members, and by their ties of warm friendship with the writer, the group has continued the tradition of the cosmic-ray group of which it was the heir.

This is not the place to tell the history of X-ray astronomy. I only would like to recall briefly some of the highlights. During the first eight years after the discovery of Sco X-1 only rockets had been available for X-ray observations. Even so, a number of new sources were discovered and other important results were obtained. In 1964 scientists at NRL achieved the first identification of an X-ray source with an optical object when, on the occasion of an eclipse of the Crab Nebula by the Moon, they showed that an X-ray source known to exist in the general vicinity of that supernova remnant was, in fact, the nebula itself. They also showed that the X-ray source had finite dimensions similar to those of the optical nebula. In 1968, after the discovery of pulsars (rotating neutron stars) by Anthony Hewish and Jocelin Ball-Burnell, and after radio and optical observations had detected a pulsar at the center of the Crab, Herbert Friedman and Hale Bradt independently found that this pulsar was a very powerful energy source, which, in the spectral range of X-rays, supplied about ten percent of the total energy emitted by the whole nebula.

Still another important achievement in the pre-satellite era was

PHYSICS IN SPACE 153

the identification, in 1966, of Sco X-1 with a very faint incon-
spicuous visible star. The identification was the result of co-
operation between two groups of X-ray astronomers (the MIT
group and the AS&E group) and two groups of optical astronomers
(at the Palomar and Tokyo Observatories). Surprisingly, it was
found that Sco X-1 emits one thousand times more energy in the
form of X-rays than in the form of light. (By contrast, the average X-
ray emission by the Sun amounts, in energy, to only one-millionth
of the total optical emission.) An essential step in the definitive
identification of Sco X-1 was the precise measurement of its
location by the so-called *modulation collimator*, an ingenious
device invented by Oda, that combines a wide field of view with a
fine angular resolution.

A new era in the study of extrasolar X-ray sources began in 1970
when the first satellite for X-ray astronomy went into orbit,
equipped with detectors designed and built by Giacconi's group at
AS&E. The launching occurred from the Italian platform 'S.
Marco', off the coast of Kenya, on the anniversary of that country's
independence. Hence the name of *UHURU* (Swahili for 'freedom')
by which the satellite became known. Because of the long observing
time it afforded, *UHURU* dramatically increased our knowledge of
the X-ray sky. Besides discovering hundreds of new X-ray sources,
it obtained many other important results, of which, perhaps, the
most striking was the discovery that some X-ray sources are binary
systems each consisting of an 'ordinary' star and a neutron star.
(The X-ray emission is powered by the gravitational energy release
in the transfer of matter from the atmosphere of the ordinary star to
the neutron star.)

Several other X-ray satellites were launched in the following
years by different groups. In 1975, Nora and I witnessed the
launching, from the same 'S. Marco' platform used for the launching of *UHURU*, of *SAS-3* an X-ray satellite designed by the MIT
group with George Clark the principal investigator. All of these
satellites produced new information which gradually enlarged our
knowledge of the complex phenomena related to the X-ray emission by different celestial objects.

A major contribution of *SAS-3* was a detailed study of the so-
called 'bursters', a most remarkable type of X-ray stars of which the

first was discovered by a Dutch satellite and many more were discovered by SAS-3 itself. In the language of X-ray astronomers, 'bursters' are sources whose X-ray emissions occur in the form of isolated 'bursts' separated by time intervals usually of hours, occasionally much longer. They have been the object of an intensive study by members of the MIT groups, particularly by Walter Lewin. The great complexity and diversity of this phenomenon denies the possibility of attempting to present here a summary at its features.

In the meantime, a consortium of scientists led by Riccardo Giacconi and supported by NASA had pushed forward a program aimed at the development of an X-ray telescope. Work on this project had been initiated at AS&E, and was then continued at the Smithsonian Astrophysical Observatory. Grazing-incidence telescopes were first used for X-ray observations of the Sun. In 1968 a team of scientists of AS&E, headed by Giuseppe Vaiana, obtained the first high-resolution X-ray image of a solar flare by means of a grazing incidence telescope carried aloft by a rocket. Other X-ray solar features were observed in later rocket flights. An X-ray telescope provided by the same team was an important component of the astronomical observatory known as *Skylab* which was sent into space in 1973 by a manned satellite. As an example of the results achieved in this mission, Figure 6.8 shows the X-ray photograph of solar corona. (Note that, in visible light, the solar corona cannot be seen against the many thousand-times brighter solar disk, except during total solar eclipses.) Eventually, in 1978, the first grazing-incidence X-ray telescope capable of extrasolar X-ray observations went into orbit aboard a satellite that became known as the *EINSTEIN Observatory*.

For much of its life, *EINSTEIN* was operated very much as a national and international orbiting observatory, available to any astronomer with a sound program suitable to the *EINSTEIN*'s observational facilities. One of these programs was a spectral analysis of X-rays by means of a high-resolution Bragg spectrometer placed at the focal plane of the telescope, which was developed at MIT by George Clark and Claude Canizares.

Because of the high sensitivity and fine resolving power of the telescope (arc seconds), *EINSTEIN* produced a vast amount of new information which was of vital importance for the progress of

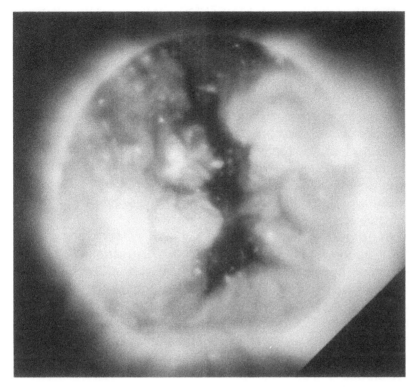

Fig. 6.8. An X-ray photograph of the solar corona. Courtesy of
Giuseppe Vaiana.

astronomy. Without going into details, I shall mention that I was
particularly interested in the results concerning X-ray sources
located outside our galaxy; first among them quasars and active
galaxies.

For some time, quasars had been known to astronomers as
powerful, point-like extragalactic sources of radio waves and light.
Before *EINSTEIN*, three quasars had been found to generate also
X-rays. *EINSTEIN*'s observations showed that practically *all*
quasars are powerful, point-like sources of X-rays.

In the language of astronomy, the term *active galaxies* denotes a
group of galaxies of different kinds, whose common feature is the
presence, at their centers, of powerful point-like sources of radiation. Several active galaxies were known to be X-ray sources

before the launching of *EINSTEIN*. X-ray emissions by many additional active galaxies were detected by *EINSTEIN*. Observations by this satellite also proved that most, if not all, of the X-ray emission of an active galaxy comes from its nucleus.

The view is gaining credit that quasars are galactic nuclei whose exceptionally high luminosity conceal the galaxies to which they belong. To discover the nature of galactic nuclei and quasars, whether or not identical objects, is one of the most challenging problems facing astrophysics. If these objects are made of ordinary matter it is exceedingly difficult to reconcile their very small size with the exceedingly high intensity of their radiation. This is the reason why the possibility is seriously taken into consideration that these objects may be super-massive black holes.

As for the role that I personally played in the development of X-ray astronomy, I must admit that it was more that of an interested spectator than an actor. For, having reached a certain age, I felt that the time had come for me to let others, of the younger generation, advance the work which, for many years, had been so much a part of my life. Besides, by then, X-ray astronomy had become a big enterprise whose progress depended on the co-operative work of hundreds of skillful scientists. I did not see the point in my attempting to make some minor contribution to their efforts. More useful, and personally more gratifying was an attempt to organize the manifold experimental and theoretical results of X-ray astronomy into a comprehensive and coherent picture, which I then might use as a basis for lectures, seminars, and articles.

This was, in essence, an educational activity, intended to acquaint the public with the extraordinary achievements of a new science. In doing so I was motivated by my interest in education which was born in the remote days of Padua and which has persisted in one way or another throughout my life. The same interest was the reason why, after my retirement from MIT, I gladly accepted an appointment at the University of Palermo, where I was attracted also by the bonds of friendship, which, through the years, I had developed with members of that institution, such as Ugo and Beatrice Palma, Giuseppe Vaiana, and Marcello Carapezza.

Lately, in writing these notes, I again had in mind an educational purpose. For I thought that the general public and, in particular, young people, might have some interest in learning first hand about the life and the work of one of the scientists of our time and to be made aware of the diverse factors, which have shaped his professional activity, factors such as rational reasoning, intuition, technological challenges of a project, and the consciousness that the riches of nature far exceed the imagination of man.

NORA LOMBROSO

As for me . . .

To Nina and Tommy

My entrance into the life of a scientist coincided with the end of the second chapter of the *Moments in the Life of a Scientist*, but my discovery and surprise at the complete 'Fusion' with his life was revealed to me in the first days after we arrived in the United States when I began receiving invitations addressed to Mrs Bruno Rossi. Thus, I realized that I had lost even my first name which I had lived with for twenty-four years.

My maiden name is very well known. In Verona, there is a monument of my grandfather, Cesare Lombroso. I remember as a child, the large crowd gathered for the inauguration around a marble figure holding a skull about which I knew only that it was my grandfather. However, with the advent of fascism, the Lombroso family fell out of favor. My two aunts, Paola Lombroso Carrara and Gina Lombroso Ferrero, both well-known writers, started to find more and more doors closed to their writing because of their anti-fascist attitude. My uncle, Mario Carrara, the successor to my grandfather's chair in Turin, spent several months in the prison where he had formerly been the doctor, while my other uncle, Guglielmo Ferrero, a historian, left Italy to teach in Geneva. My father, instead, ended up as a professor of physiology in Sicily, a region which Mussolini considered (along with southern Italy) as a kind of colony in which to keep the 'dangerous men' at a distance. Thus, I suckled anti-fascism with my first milk. I absorbed it, at first of course without understanding, but soon it became a vital part of my life and attitude.

However my life and activities is another story, or maybe not; it is also the story of 'Mrs Bruno Rossi' as I was not born the wife of a scientist. I became one, carrying with me my past twenty-four years, my ten years of life in beautiful Sicily, my admiration for my family and the stand they took against fascism, my friends, dead in Spain or shut in prison, my passion for art, and the thousand other things that remain hidden inside me, but are always present.

I remember very little of Padova, the city and the house where Bruno and I made our first home. Perhaps it was because the times (1938) were so uncertain and the deteriorating political situation in Europe and in Italy placed an ever increasing tension on my everyday life. What I do remember is my departure, the rushed goodbye to 'my' Giotto in the Cappella degli Scrovegni and the rage I felt against Bruno, who did not want to leave 'his' laboratory, which was no longer his. For him, it was like a cord he could not cut.

A month before, on the fifteenth of September, the anti-semitic laws had come out. The event was startling news, not only for the Jews, but for many other Italians. The campaign started by Mussolini under pressure from his friend (or boss) Hitler had been going on for some time, but the Italians had responded with lukewarm enthusiasm. As a matter of fact, the whole issue of Judaism and Christianity was more of a religious question than a racial one; you converted – you became Christian. There might have been some social snobbery, but I never was aware of it. The whole idea of an Aryan race was often laughed at. I remember the thousands of Jews fleeing from Germany, Austria, Yugoslavia, etc. that had been for years welcomed and, later, sheltered all over Italy. I remember our friends (or some of them – the brave ones) rushing to our house totally in shock at what was happening to us.

We decided to leave, but I did not have a passport to go abroad. Right away I thought of my friends in Sicily and wrote, I don't remember to whom, but only that I wrote 'I have to leave Italy. Bruno can not work here anymore and we have to depart. Search for a cousin, a friend at the ministries who can give me a passport'. I still don't know who carried it off, maybe many, but the passport arrived and we were ready to leave. The last day in Padova I joined Bruno in the laboratory in order to pull him away. I remember the wide staircase as we, melancholy, descended slowly. At the bottom of the stairs stood Mario, the janitor, in tears. 'Professor, don't leave. Why? Why? It's not fair. It's not fair.' It was the most moving goodbye that we received from the people of our native land.

I had been married for only a few months. In August, I had met many physicist friends of Bruno while vacationing in the Alps:

Fermi, Bernadini, Rasetti, Amaldi, and others. I felt confused and insignificant in their presence. No one was speaking of politics as I was used to in the house of my father; they were all cheerful and amusing, except when they started playing with mathematical puzzles and riddles which were incomprehensible to me. The mountains were magnificent and perhaps for this reason Padova seemed so gray and pale.

I don't remember the goodbyes with our families. We were going to Denmark. It was nearby and it was not clear then for how long we would be gone. Besides, the consequences that Hitler's invasion would have on the fate of the Italian Jews and on all Italians was not clear in its monstrosity.

Denmark received us with a grace and warmth I will never forget. Mrs Niels Bohr had the gift of speaking to you as if she had known you forever, finding the right words and warm gestures which made you feel at home and gave you courage. I remember her, a beautiful hostess in their beautiful house, where we were received so often like old friends. I was sorry to leave Copenhagen, lovely, airy and modern, for the fogs of Manchester. But, I liked to travel and loved to get to know new people.

I met the Bohrs much later, in of all places Palermo, where I was visiting for a few days. Mr and Mrs Bohr were there on vacation and stopped by the Physics department of the university to see what was going on. They were met by the janitor who told them that everyone was in a faculty meeting upstairs. After some time he mentioned to Beatrice Palma that an old German couple was downstairs waiting to see them and still later, Beatrice went down the hall and discovered the two Bohrs sitting patiently on the only hard bench in the entrance hall. She was in shock. Hurriedly, she and all the physicists and friends, including me, organized a big party where we all had fun together with the 'old German couple'.

In Manchester we found the Blacketts. What a joy to be able to speak Italian with Constance after all our efforts in Denmark to make ourselves understood in English or German. After a brief stay in a pension decorated with framed paper hearts and with paper flowers everywhere, eating cabbage, potatoes and mutton, Patrick dug up a nice little flat in a never-ending street where three steps and identical doors led to identical apartments. Actually, to three flats.

We lived on the first floor off the common entrance for the three flats. Unfortunately, the only telephone in the building was also on the first floor and I became the telephone receptionist; I with my primitive English! Maybe that is why, even today, I still hate the telephone.

English was a problem, but not the English. They were very kind, positive and ready to help us. I remember the efforts of everybody on the bus when I was asking for Worthington Road (my street), as they tried to interpret what in the devil I was trying to say.

I met Occhialini in Manchester for the first time and he would lighten our misty days with his spark. I learned to cook, so-to-speak, a hard job as Bruno did not want to touch a leaf of cabbage and not many other vegetables were available. In Manchester I also learned that the wife of a scientist, one with a law degree and a passion for painting, must learn to be a housewife, no easy task, even in a house as small as ours, with only a fireplace for heating and a freezing bathroom.

Patrick Blackett was our first guest for supper. I bought a beautiful and expensive roast which I had never prepared in my life. Bruno approached it to cut it in proper English style with the most impressive knife from our kitchen. But, as soon as he touched it, the roast flew elegantly off the plate, ending up on the floor. The three of us were left speechless. Patrick was the first to recover, 'Well Nora, pick it up and we will try to sharpen this knife'.

My culinary experiences were many and often disastrous, but I remember more vividly our bicycle trips in the countryside around Manchester, the moors, the early delightful spring, the trips to London and Cambridge, the visits to the museums and the mouth-watering dinners at Bhabha. I remember the anxious discussions at the Blacketts about the political situation as it became ever more tragic in Europe. In London, we were taken under the wing of the Crowthers. He was a great journalist at the *Manchester Guardian* and his wife an anti-Nazi German. Both were so very interesting and kind. Their house was flooded with refugees, at that time especially Spaniards; poor souls, they were even more lost than ourselves.

Refugees were coming to England from everywhere: Spaniards, Czechs, Hungarians, etc. One day, Mrs Blackett asked me to stand guard over a painting that Picasso had sent to England for an exhibition to collect money for the Spanish refugees. Thus, I spent hour after hour in front of *Guernica*, for me an artistic revelation and vivid expression of the situation in which we were immersed.

Still, life in Manchester proceeded relatively quietly. Bruno was working with instruments that mostly functioned by themselves and so he was free enough to ramble about the countryside. The English weekends were sacred; the doors of the Laboratory closed promptly at five.

However, our wanderings around the globe had just started. Everyone in England decided that the situation there was getting to the point where it was safer for us if we left Europe. This was a hard blow for me. I could not accept running away from the European tragedies in which I was so deeply involved. When the decision was made, we went to Geneva in 1939 to say goodbye to our families. This time we all felt that it was going to be a more definite separation. As a matter of fact, it was to last nine years.

The *Liberté* was a handsome ship, packed with cheery tourists on visits to America, a contrast to our troubled mood. New York was beautiful in the dawn, wrapped in a light mist with all the sky-scrapers welcoming us through the haze. I did not notice the Statue of Liberty; probably I did not know its meaning. The language that I had barely learned in Manchester was of no use here. Here they called 'minced meat', ground beef, 'tins' of peas were cans, and Columbus Circle was not a circle, just a crossing of streets. Laura Fermi was not very encouraging, 'Remember Nora that even though Bruno is rather well known among physicists, you will always be refugees here'.

In New York, I first saw a black woman in person. I was straightening up my 'service' apartment when she appeared, a big woman as I'd seen in the movies, she mumbled something and forced her way into my room with a broom and a lot of paraphernalia. Only then did I understand what 'service' apartment meant and I was delighted.

Soon we left New York for Chicago, our supposed 'destination'. It was a new world to explore, so different and far from Europe, as if I had landed on the Moon. It became easier for me to break my ties to Europe, the old world. There, I truly became 'Mrs Bruno Rossi' as I understood that our life and means of survival depended solely on the success of Bruno in his career in a period when the American depression was still very real in all the universities.

In Chicago, we were sponsored by Arthur Compton. The Comptons invited us to their cottage in Wisconsin, a very wild and beautiful place where I learned to clean fish fresh from the pond. But our vacation only lasted a short time as Bruno, with his bright ideas on experimental projects, inflamed the organizational spirit of Compton. Soon he sent us back to Chicago to organize an expedition to the Rocky Mountains in Colorado to bring Bruno's dreams to life. I also took part in the success of this expedition: blowing into glass tubes, so hard to clean, soldering miniscule wires, and preparing quick meals. At the same time, one of Bruno's collaborators was building a huge machine with hundreds of small lights that was quite impressive next to the wooden boxes and glass tubes of Bruno's experiment. The working rhythm was very intense; day after day passed with no relaxed dinners and no chatter, as if life depended on those strange instruments. It was a bit frightening. Bruno worried that he could not make it in this intense atmosphere, so full of frenetic activity. In the middle of August we left on one of the most dilapidated busses I had ever seen and crossed the Midwest to Denver. How long the trip took I don't remember; what sticks in my mind were the infinite seas of wheat and corn, red seas of tomatoes and huge trees making dark spots on the long straight highways; boring but fascinating. I believe, still today, that no one can understand the United States if he has not crossed the country by car, though I would advise a machine a bit more comfortable than our old bus.

In Denver, we were the guests of the Stearns, friends of the Comptons. They taught us the rules of baseball and I showed them how to prepare spaghetti. Very soon we started the climb to Mount Evans. Autumn was near and Bruno was afraid to be caught by the snow. When we made it to the top, we discovered that the cabin where we were to sleep was way up on top of a rocky hill that we had to climb on foot. Bruno and I took a blanket, some water, and a can of food and started up the hill towards our refuge. Our two companions, both big fellows, tall and strong, were in heaven at that altitude (1400 feet) and to our surprise, started to carry up all of our provisions (these Americans!). I spent the first night nursing our friends, downed with a serious case of mountain sickness – they were from the plains of the Midwest and had never seen a mountain. The next morning, Bruno and I returned down alone to the saddle where we had left the bus and started to set up the very heavy pieces of lead above and below the geiger counters. Later I was asked to sit and count the tic-tac of the cosmic rays and only then did I accept for the first time the existence of cosmic rays. Until then I had been convinced that they were an invention of Bruno and other scientists.

In Echo Lake we cooked in a log cabin where, on the first night, the squirrels ate all the supplies stored inside and the camp birds finished the fruit left for a minute outside. The lake was lovely, surrounded by thick forests and high mountains reflecting in the clear water. Cooking in the cabin was not easy; spaghetti disintegrated without cooking and nothing worked in the usual way. None of the scientists had explained to me that at 4000 meters, water boils at 80 degrees!

The experiment was finished barely in time as the trees turned a bright gold – not on top as we were above the timber line, but down the mountains all around us. We took the bus and returned to Chicago. Our furniture from Italy was awaiting us in a big crate. My mother had sent us everything she thought we might need to survive in the new world, paying for it with the money that we were not allowed to take out. We found a pleasant attic, but soon discovered that part of our furniture could not make it up the narrow stairs. I remember the boys from the moving company, two Italo-Americans, all excited to see the paintings and objects from Italy, a land that for them was magical and unknown.

In Chicago, everyone welcomed us with singular warmth. Flipping through the pages of a calendar from that period, I can see that we were invited out or I had guests at home almost every evening. Giuseppe Borgese was teaching Italian literature at the University of Chicago. He had left Italy together with Salvemini, Ferrero, and many others after the establishment of fascism. He was the only Italian intellectual that we met in Chicago and he explained that we would be considered strange beings there as, for Americans, Italians were, at that time, all well diggers or bricklayers or shoemakers – *never* intellectuals. But, strange or not, we soon felt at home and made many friends. Alas, because of our many moves and enormous distances in this country, we soon lost contact with them. This is very sad for me; I would like to hold on to every friendship and every experience. Instead, my life has been a kaleidoscope that turns rapidly, never returning.

The Comptons lived in a beautiful house. He was as handsome as an actor, very 'simpatico' and kind, and soon became very attached to Bruno. Mrs Compton and I also became very close and she brought me on many Sundays to her church, hoping, I believe, to make a new convert. The music was very beautiful and so was the singing, but I could not make out what was going on and remained unconvinced. Many years later I set myself to study with interest the multitude of protestant sects, until then unknown to me.

In the autumn of 1939, Hideki Yukawa passed through Chicago and came to supper. We could communicate only a little then, with our Italian English and his Japanese English. Only later, after the war, did we get to know Yukawa much better. Once when he and his wife were our guests, Mrs Yukawa performed several lovely dances in kimono and fans on December 7, 1946, at the sixth birthday of our daughter, leaving Florence and her school mates enchanted. Still later, when we visited Japan, we met them again at a banquet that they gave for us in an elegant restaurant.

We saw the Comptons many times after we left Chicago, also in Cambridge where we moved after the war, where another Compton, Carl, his brother, was the president of MIT. They are both dead now, as are many of our protectors and friends.

It was the end of the great depression in the United States and Bruno's position in Chicago was very insecure and the salary low. Therefore, we visited several universities by train and bus: Notre Dame, University of Wisconsin and others, hoping to find a more stable job. I don't remember being very worried, maybe because I was young, perhaps just naive. Saving pennies, we were able to buy a huge secondhand car and I was expecting our first child when we embarked in the spring in our car to cross halfway across America to reach Washington and attend a conference. Bruno had had all of ten driving lessons in Padova and had never owned a car. And so we departed, with all our friends terrified for our safety, having seen us on the streets of Chicago. They hurriedly gave us some advice: if there is smoke coming out of the motor, check to see if the brake is on; always try to back into parking places; and absolutely do not go faster than forty-five miles an hour on the highway. I remember a thousand frightening encounters before we reached Ithaca in the middle of a snowstorm. Bruno had been invited by Bethe to give a lecture at Cornell and I was offered tea by the wives of the Physics department where I was also scrutinized. Only later did I understand that Bethe, who had great affection for Bruno, had made the invitation possible hoping that Bruno would be offered a chair. We left for Washington where the cherry blossoms in full bloom welcomed us – what a show! There I met my brother who had just arrived from Italy.

On our return to Chicago, Bruno threw himself back into the frenetic work in preparation for a second expedition to Colorado. A pleasant interruption during the summer were the lunches at my home with Fermi who was teaching at the University for the summer and living in student housing where he said the food was only good for dogs. Bruno and Enrico chatted while I cooked a few meters away. Fermi kept mentioning some incredible experiments on atoms, nuclei and other things incomprehensible to me. Only later, in 1943, in Ithaca when Bruno, very worried, told me that perhaps we would have to leave Cornell to go to a mysterious place, and I saw his troubled face, did I remember Fermi speaking of the possibility of a monstrous bomb emerging from those experiments they had discussed. 'Bruno, is it possible that you are going to work on that bomb that Fermi mentioned?', I asked. Bruno turned pale, 'How do you know about the bomb? It is a great secret. No one must know anything. What are you saying?' So I understood and kept tight inside me this terrible secret which I happened on involuntarily.

In August, we left for Colorado. I was several months pregnant and the doctor had advised against going up and down the mountain. So I remained in a small room in Denver under the protection of two elderly landlords who filled my room with magazines so that I would not get bored. By now we knew that we had to once more pick up stakes and move to Ithaca. I was sorry to leave the interesting and warm environment at the University of Chicago although I hoped to find less snow and ice in the new location.

We arrived in Ithaca in the late autumn of 1940. They had found a lovely small cottage with three pines in front, a green, green lawn, and large sunflowers. Until then I had not seen the typical American small town and, in October, it appeared very beautiful. However, by the seventh of December, when the snow and ice was everywhere and poor Bruno had to take me by car to the hospital several miles away, I changed my mind somewhat. At any rate, we made it, and Florence was born in the hospital and not in the car. I returned home with a pretty and hungry child and my life changed somewhat. No more typewriting Bruno's articles or following his work, but instead diapers and 'pablum'. The most difficult period for me was when Bruno started going to Boston to work on the radar. The war was in full swing and we were without news from our families; soon Roosevelt rationed gasoline and I found myself even more isolated than ever. Still I felt lucky when I thought about what was happening in Europe.

On December 7, 1941 some friends came for supper and were surprised to find us in the dark about the event of the day, the attack on Pearl Harbor. We had been too busy in the kitchen preparing a cake for Florence's first birthday. It was not a surprise for us; we had felt at war already for a long time and America's entrance, although in a way so tragic and unimaginable, seemed inevitable. The news from Europe was ever more frightening and we thought the entrance into the conflict of this great country could change the situation. Meanwhile, Bruno continued to work in Boston and I did not know what he was doing.

When Florence was twenty-two months old, the Chairman of the department of Languages knocked on my door and asked if I was willing to teach conversational Italian. As they were in a hurry to find a native Italian to teach future soldiers the rudiments of the languages of the countries with whom they were at war, I was asked to respond within twenty-four hours. I found myself torn between the problem of how to care for my child and everything else, and my desire to contribute something to the war effort. The university's nursery school helped by accepting Florence even though she was under the age limit and I went to buy myself a suit in order to acquire the professorial look, if not the substance. I began to work feverishly with the only grammar book I could find in Ithaca and I taught all that year until Bruno announced our impending departure. Hans and Rose Bethe, our dearest friends, had already left and I did not know for where. I understood that we were going to join them. Bruno was in a very bad mood; I too, at having to quit my opportunity to teach, so the Rossi household mood was rather stormy for several weeks. Bruno left for the mysterious place. Our furniture went back into storage after only a few years outside; I sent a few paintings and some favorite objects and so we left for the unknown destination which turned out to be Los Alamos.

Los Alamos had been described to me as a small town of barracks on top of a mountain. As we drove on a winding road, climbing up from the Rio Grand, a city appeared in the distance, full of mysterious lights, scattered on a strange plateau – a mesa – surrounded by a new and marvelous landscape. The barracks turned out to be two-floor wooden houses with four apartments, pleasant enough. From my apartment I could enjoy a beautiful view of trees and a narrow gorge that ended in a wild valley which was a joy to all the neighboring children. However later, much later, bulldozers arrived and in a few hours leveled everything. No more gorge, no more trees, everything disappeared.

I soon made friends with my neighbors and Florence found a great number of playmates. Almost everyone in Los Alamos worked at some kind of activity connected to the laboratory or the community. After a short period at the microscope observing mysterious dots, I was asked to be a 'gal Friday' for Otto Frisch, the well-known Austrian physicist whom I had met in Copenhagen, and I worked for him until the birth of my son Frank.

The landscape around Los Alamos is one of the most beautiful in the world. The contrast between the desert and luxuriant mountains, between the white winters, the spring all in bloom and the summer, dry and golden, left me with many nostalgic memories. Many Indians, men and women, came from the nearby pueblos to work at the mesa; the women in their long colorful skirts and long black hair, dignified and reserved. For me, worried about living in the exclusive environment of superscientists, getting to know the Indian population, known to me previously only through Westerns, was a very special experience. Altogether, life in Los Alamos was pleasant, without the rivalry between scientists that we had expected to find. For this, I give full credit to Oppenheimer's intelligence and organizational ability. The Oppenheimers were a very complex couple and even though we were invited often for supper at their small house decorated with French Impressionist paintings at that time I did not really get to know them well. Later, we saw more of Oppenheimer when, after the war, he came to visit us in Cambridge and tried to convince Bruno to leave MIT and go to Princeton where he was teaching.

In Los Alamos, Bruno was working in a gorge very far away where he was engaged in experiments I knew were dangerous. Sometimes he had to rush out in the middle of the night down winding roads in decrepit cars. Once, the guards almost shot him when the car, having lost its brakes, passed the guard post without stopping. In fact, we were surrounded by barbed wire and had to carry identification cards with us. Once when we went on a weekend vacation with some friends to a ranch, we had to tell everyone we lived in Santa Fè (Los Alamos being a secret). Meanwhile, the children had invented a marvelous game of stopping everyone and demanding that they show their ID cards!

On November 10, 1944, our son Frank was born. He was big and handsome, but after two days the pediatrician told me that he had an intestinal occlusion. Our small hospital did not have a surgeon and only after what seemed centuries did one arrive at the mesa and operate on Frank. He survived the operation, but my life changed as Frank remained rather delicate. A young Indian woman came to help me and became so attached to Frank and I that when we were to leave Los Alamos she asked to come with us to Boston. But how would an Indian from New Mexico find herself in Boston, the city of the Yankees and Paul Revere? I did not feel right about having her come with us. I never would have dreamed at the time that my daughter Florence would, many years later, give me a handsome grandson with Indian blood.

In February 1946 we left New Mexico. We arrived in Boston in the middle of a heavy snowstorm: children, furniture, and a new house in the suburbs. This happened forty years ago but I cannot believe it. My friend Carlo Levi, the writer, once told me when I was younger,

'You will see, Nora, the older you get, the faster time goes by'. Now I am older and he has left us for good.

Four young collaborators of Bruno's in Los Alamos joined us in Boston. They became the 'four musketeers': Sands, Bridge, Thompson, and Williams, the nucleus of a crowd of students and collaborators that arrived at MIT from all over the world. The 'Cosmic Ray family' was a tight, friendly group that invaded Bruno's working days and also our house. Eventually, we bought a large and beautiful home in Cambridge. It soon became the center of comings and goings of guests, of parties and of adventures, some comic, some sad, that filled my days. Cambridge is a special sort of town. Harvard dominates, students invade every corner, big shot professors create a very rarified atmosphere that is hard to ignore. Everything was so different from my previous experiences in America, and I missed New Mexico and the mountains (and Italy with its old stones). So, after a half-hearted attempt to get a Masters (refused because 'with two small children, you can't occupy a bench that a young girl would use to better advantage'), I decided to throw away my intellectual ambitions and occupy myself with practical, simple and fun jobs. I was a cook in a small restaurant, a sales person in nice galleries, an importer of Italian crafts, a puppeteer, a translator, and a teacher. At the same time, with Bruno, I welcomed many visitors from all over the world: France, Italy, Japan, India, Australia, South America, etc., etc. Poor children of mine, exposed to so many cultures while I was attempting to make them into 'real Americans'. Still, we made many good friends from all over the world.

I remember many springs when with the blooming of the first flowers Vickram Sarabhai would arrive and brighten our days and evenings with his enthusiasm, vivacity, and sex appeal. And Minoru Oda, very young, catapulted into Boston after a long trip from Japan by boat, his broken English, his marvelous drawings that left my children open-mouthed in admiration, the hundreds of squid fished by him that filled every container in my cabin in Cape Cod. And Matt Sands who came to pick us up on our arrival from New Mexico and got lost in the neverending Mass. Avenue, a man so young, but who had already held every job I could imagine, and who was my savior any time something broke in the kitchen, until he left abruptly to go west. And John Tinlot and Bernard Gregory, both so gifted and pleasant, who died so very young, leaving a great void among the people who had known them.

One day young Yash Pal arrived from India, shortly after a terrible hurricane had felled some tall trees in our yard. He appeared, all dressed in white, while a number of the cosmic-ray family were intently sawing away at the trees and eating lasagna. I hastened to reassure him that Bruno did not normally expect his students to saw fallen oak trees.

As soon as we could we took long trips. And so I lived for a few months in Japan with Tomoe and Minoru Oda, in India with the Sarabhais, and in Bolivia with the Escobars. In Kenya, after seeing a satellite launching, our dear friend, George Clark took us for a tour of the savannahs. It was an unforgettable experience to see the thousands of handsome animals moving freely around us, almost surprised at our intrusion into their land. The little Alfa Sud, however, broke down for the second time just after we had passed a large pride of lions. As Bruno and George nervously got out of the car (I should also mention that we had also passed a sign saying that under no circumstances should you *ever* get out of the car!) to try and figure out what was wrong, I looked through the back window to watch for dangerous animals. I saw with joy another car approaching driven by a native who rapidly fixed our car.

Of course, very soon after the war, we went back to Italy for the first of many visits to see our families and friends. In 1947, Italy was still all in ruins, but everybody was frenetically working at healing the ravages of war, repairing buildings, bridges, and restoring the thread of their lives. In 1953, our third child arrived together with a poodle puppy who became an integral part of our family. Bruno was always very busy. His work at MIT was not enough for him and with two of his students, Martin Annis and George Clark, he founded a company (everybody was saying three crazy cosmic-ray scientists). I enjoyed very much to see the beginning, the development and success of an 'American business'.

After many years in the United States, I still have a strong Italian accent, but when people ask me 'where do you come from', I resist being put in a cubby hole and answer 'from New Mexico' and if they insist before, 'from Ithaca', and still before, 'from Chicago', and only then do I respond to their curiosity by admitting that my accent is Italian and, finally satisfied, they say, 'you have such a nice accent'.

I think back sometimes to all those years gone by so fast. Was I a scientist's wife, or instead just a wife, a mother, a grandmother blessed by two beautiful grandchildren, or maybe even simply, an immigrant woman. I did not pass through Ellis Island, I had it easier. I belong to two homes, the USA and Italy, but I still wish that there was no Atlantic Ocean, no distances so great, no parting of friends.

Let me finish with an episode which happened some years ago. It was Thanksgiving day and I had gone on strike, refusing to cook the traditional turkey for Bruno and our children. And so, while we were sitting in a restaurant, a young man approached our table and offered us the end of his bottle of wine. Turning to Bruno, he asked, 'I made a bet with my friends. By any chance are you a poet or an astronomer?' To which I replied, 'He is both'. But I was so surprised that this double aspect of Bruno's personality was impressed on his face.

INDEX

180   INDEX